Vehicle Simulation

Vehicle Simulation
Perceptual Fidelity in the Design of Virtual Environments

by

Alfred T. Lee

CRC Press
Taylor & Francis Group
Boca Raton London New York

CRC Press is an imprint of the
Taylor & Francis Group, an **informa** business

CRC Press
Taylor & Francis Group
6000 Broken Sound Parkway NW, Suite 300
Boca Raton, FL 33487-2742

© 2018 by Taylor & Francis Group, LLC
CRC Press is an imprint of Taylor & Francis Group, an Informa business

No claim to original U.S. Government works

Printed on acid-free paper

International Standard Book Number-13: 978-1-1380-9452-9 (Hardback)

Library of Congress Cataloging-in-Publication Data

Names: Lee, Alfred T., author.
Title: Vehicle simulation : perceptual fidelity in the design of virtual environments / Alfred T. Lee.
Description: Boca Raton : Taylor & Francis, CRC Press, 2017. | Includes bibliogaphical references and index.
Identifiers: LCCN 2017028058| ISBN 9781138094529 (hardback : alk. paper) | ISBN 9781315105987 (ebook)
Subjects: LCSH: Automobile driving simulators. | Virtual reality.
Classification: LCC TL152.7.D7 L44 2017 | DDC 629.04/6078--dc23
LC record available at https://lccn.loc.gov/2017028058

Visit the Taylor & Francis Web site at
http://www.taylorandfrancis.com

and the CRC Press Web site at
http://www.crcpress.com

Contents

Preface

The term *virtual environments* applies to those synthetic or artificial environments that attempt to recreate real-world environments and the devices that an individual or group might use in those environments. Virtual environments include simulators such as those used by drivers, pilots, police, surgical, and a variety of other disciplines. The history of the development of these devices is replete with examples of applying the latest technologies in an effort to replicate real-world environments. One form of virtual environment, flight simulators, has perhaps the longest history of development and contains many examples of this attempt to achieve complete physical replication or what in this book is termed as *physical fidelity*. The earliest example of flight simulation that achieved some commercial success was the Link trainer developed in the late 1920s. The Link saw extensive service in pilot training before and during World War II. The trainer is an example of how the technology available at the time was used in an attempt to replicate the appearance and function of an aircraft cockpit. Although it contained aircraft instruments and a primitive motion platform system, the device had no visual image of the external environment. The pilot was completely enclosed within the cockpit and the trainer even had aircraft appendages such as wings and a tail assembly. The outside of the trainer was painted blue in what appears to be a final attempt to make it appear similar to a real aircraft. With the pilot fully enclosed in a cockpit with no external reference, the Link trainer represents one of the earliest examples of total immersion simulation. The Link trainer is one of the most successful of the early simulation devices as it was used to train thousands of pilots in instrument procedures. The crude motion platform system was, however, later abandoned as ineffective.

Edwin Link, the designer of the Link trainer, appears to have followed a simple dictum: if you want to simulate something in the real world, make the simulation appear and behave as much similar to the real thing as possible. In the decades that followed, all virtual environments in training, testing, and research have, more or less, followed this dictum. This dedication to physical fidelity in design has some obvious benefits as well as costs, which will be discussed throughout this book.

This book argues for a very different design goal for virtual environments, that is, one based on perceptual fidelity and not on physical replication. Perceptual fidelity emphasizes the importance of producing human behavior, specifically perceptual and perceptual-motor behavior, in virtual environments comparable to that found in real-world environments under similar conditions. It is logical that perceptual, rather than physical, fidelity should be the design goal as it is human behavior that is of interest in most application areas of virtual environments.

The differences between physical and perceptual fidelity and the ramifications of these different goals on device design are explored in Chapter 1 of this book. Chapters 2 through 5 deal with the general human perceptual processes and with perception in simulated environments. The emphasis on vehicle simulation in this book reflects not only the importance of this area but also the availability of a large database of research. Chapters 6 and 7 address the problem of quantifying perceptual fidelity and the impact of perceptual fidelity on the design and evaluation process. These two final chapters are intended to have broad applications to virtual environment design beyond that of vehicle simulators.

Although Chapter 1 presented the arguments supporting the implementation of perceptual fidelity in vehicle simulator design, Chapters 2 through 4 address the role of perception specifically in vehicle operations. These chapters correspond respectively to the three major areas of perception and perceptual-motor skill that affect the vehicle operator performance. They are vision, perception of physical motion, and force perception.

Chapters 2 through 7 are intended as both a review of what is known about perception in vehicle simulators from the past research and how perceptual fidelity can be incorporated in the design process from the beginning. Quantification of perceptual fidelity is also discussed at length and how the level of perceptual fidelity in a device design can be measured.

Chapter 2 describes the fundamentals of human vision, the basic processes of visual perception, and the visual imagery systems that are used in vehicle simulators to recreate the visual experience of vehicle operation. The basic neuroanatomy of the human visual system and the basic capabilities and limits of human visual perception are described. The effects of variations in visual imagery system design on human–vehicle performance are addressed.

In Chapter 3, the perception of physical motion is described, including the mechanics of the vestibular system. Basic capabilities of humans to detect changes in velocity are identified, including response profiles to angular and linear accelerations. The technology of motion platform systems and related technologies are reviewed as are the limitations that these technologies place on the simulation of vehicle movement. Reviews of data on the effectiveness of motion platform systems and possible alternative motion cueing designs are also discussed.

Force feel and force feedback are essential to vehicle control, but are poorly understood due to the low priority that these cues often have in vehicle simulator design. Chapter 4 addresses the issue of the perception of force feel and feedback with reference to the two key sensory systems: (1) the tactile and (2) proprioception systems. Both of these systems are part of a larger system of body sensing called the somatosensory system. This system is described and the basic capabilities of touch and force perception are discussed. The effectiveness of current force forward systems is reviewed.

Chapter 5 addresses the less well-understood issue of how sensory systems combine, as they often do, in response to real-world stimuli. Multimodal or multisensory perception involves two or more sensory systems, such as vision and hearing in speech perception. In vehicle operation, the most common multimodal perception occurs in vehicle movement in which both vision and vestibular sensory systems are engaged. Examples of specialized multimodal perceptual systems such as postural control and their relevance to vehicle control is described. Specific issues with regard to the simulation of multimodal stimuli such as the problem of timing and congruence are discussed.

Chapters 6 and 7 address the problem of quantifying perceptual fidelity and how it affects the simulator design process. Chapter 6 puts forward a methodology by which perceptual cue contributions to the perceptual tasks performed by vehicle operators can be quantified. Applications are shown for both unimodal stimuli such as the visual perception of distance and for multimodal stimuli such as the combined visual and vestibular stimuli in the perception of self-motion. The development of an index of perceptual fidelity is described, which can be used to quantify the contribution of perceptual cues to vehicle operator tasks.

In the final chapter, Chapter 7, the implications of perceptual fidelity for the simulator design process is described. Perceptual fidelity is a process that begins with the definition of tasks that vehicle operators will perform with the device. The task requirement documents describe in detail the tasks that the device must support and the criterion for successful performance in each task. This is followed by task decomposition, which identifies any subtasks and all task elements. The goal of the decomposition process is to identify the perceptual cues supporting each task and subtask. The development and the role of the index of perceptual fidelity in decisions regarding the final similar design are then described.

This book is written as an introduction and explication of the topic of perceptual fidelity in virtual environment design. It reviews as much of the relevant research as possible within this framework. Other texts, including the author's earlier book on flight simulation, as well as a variety of existing edited collections of articles on virtual environment design should be consulted for more in-depth analyses of specific design issues in the development of virtual environments.

Alfred T. Lee
Beta Research

About the Author

Dr. Alfred T. Lee has been involved in the research and development of vehicle simulators for more than 35 years and is the author of more than three dozen articles in this field. He is also the author of the 2005 book Flight Simulation: Virtual Environments in Aviation. He is currently the president and principal scientist with Beta Research of Los Gatos, CA, a company which conducts research and development in virtual environments. Formerly, he was a senior research psychologist at NASA–Ames Research Center, responsible for initiating, planning, and conducting research programs in human-systems integration, including the development of flight simulators.

1 The Problem of Fidelity

INTRODUCTION

The attempt to create synthetic environments, now commonly referred to as virtual environments, has a history dating back to the days long before the arrival of modern computers. Compelled by the need for alternatives to real-world training, testing, and research environments, engineers from a variety of disciplines set forth to develop environments that would provide the opportunity for training and testing individuals under conditions that were either too hazardous or inaccessible in order to determine how they would respond. These virtual environments would also allow the opportunity for systematic scientific investigations, allowing levels of control that are unattainable in the real world.

Among the first of these synthetic environments was the Link trainer. Developed in the early 1930s, the Link trainer was one of the first full-scale attempts to create a flight simulator training environment. Despite its relatively crude technology when compared to today's modern flight simulators, it served as one of the principal training devices for pilot training in World War II. By completely enclosing the pilot within the simulated aircraft cockpit, the Link trainer was one of the first virtual environments that attempted to achieve complete immersion of a pilot into a synthetic environment. The term *immersion* has become common in descriptions of virtual environments. It refers to a psychological state in which the individual is unable to distinguish between the virtual environment and the real one. The individual is, psychologically and physically, *immersed* in the synthetic environment.

The Link trainer was developed long before the availability of modern computer technology. Not until decades later would engineers have available the kind of technologies that would make synthetic environments truly immersive. Virtual environments exist in many forms but their essential attribute is to create an environment that envelops the user of the device with sensory stimuli intended to produce a particular experience. In the case of simulators, the intent is to recreate a specific real-world experience. The term *simulator* means that the environment is intended to be the same as, or similar to, an environment in the real world. A vehicle simulator is an attempt to recreate the vehicle, the vehicle's behavior, the vehicle's contents, and the external operating environment of the vehicle.

Virtual environments do not need to be, like simulators, an attempt to recreate a real-world reality. Some virtual environments, particularly those in the gaming industry, have very little to do with reality. In fact, the synthetic environment is intended to provide the device user with a very unreal environment. A variety of these games and other entertainment media are solely intended to immerse the user in a fantasy world. The purpose of this world is to engage the user in a kind of unreality that will provide them with an experience unlike anything they might experience on this Earth.

Virtual environments are often thought of as synonymous with virtual reality where in fact virtual reality (VR) is a technological subset of virtual environments. The term *virtual reality* was a term applied by Jaron Lanier in the 1980s to a set of technologies intended to fully immerse the user in a virtual world. The typical VR consists of a helmet-mounted display (HMD) to display the virtual, computer-generated world, a data glove, or some other means of interacting with that world and an audio device such as a headset to provide audio stimuli. Once again, immersion in the form of a computer-generated world is intended to replace the real world for the device user. The VR is, therefore, an attempt to create a world for the user, which is wholly synthetic and detached from the world that the user is currently experiencing. The humble origins of the VR technology, however, made it difficult to create those synthetic experiences. Even today, VR technologies, particularly the HMD, have struggled to achieve the user experience levels that were originally promised.

Vehicle Simulators

One class of virtual environments, vehicle simulators, has been much more successful in delivering the promise that was shown with the early Link trainer. The advances in vehicle simulation technology, particularly in driving and flight simulators, have been dramatic. The past 40 years have shown widespread applications in this area in both ground and aircraft vehicles. This has been particularly true in the area flight simulators for both civil and military applications (Lee, 2005). The technological developments in this area have been so successful that some pilots receive all of their training for a specific aircraft in a flight simulator and never receive training in the aircraft itself. Military applications in flight simulator training now include advanced combat training in realistic threat environments. The latter is particularly valuable because it allows the training of aircraft pilots in the conditions of air combat without exposing the crew or the aircraft to its hazards.

Ground vehicle simulation such as those for passenger cars and trucks as well as military vehicles has proceeded somewhat less rapidly. This is partly due to the fact that training and evaluation in the actual vehicle can be done with substantially less cost than that associated with flight training. The slower rate of development of driving simulators is also due to the slower progress in visual imagery systems. Only with the advent of modern computer-generated visual imagery over a large field of view, could the vision-intensive world of driving be realized.

The advances in both computer processing and in visual imagery systems are largely responsible for the today's state of vehicle simulation technology. As supporting technologies began to reach maturity, the question of how these technologies should be most effectively used has arisen. Although the original application of the technologies in civil and military aviation could often justify the very high price of flight simulators, there is an increasing desire to address the very high cost of their development. The cost of full-mission flight simulators now ranges between 10 and 15 million dollars for large aircraft. This high cost restricts the access to these devices to relatively few applications. Very few pilots in the civilian sector will ever have access to a full-mission simulator of the type used in either commercial or in military aviation. Most of these pilots will continue to receive their training in the actual aircraft. This restricts their training to areas of flight that are deemed relatively safe for both the pilot and the aircraft. It virtually eliminates the use of simulators for training in the identification of hazardous and potentially fatal flight conditions. The first exposure to these conditions will, unfortunately, occur in the real aircraft exposing both the pilot and the passengers to real hazards.

The limitations on access to vehicle simulators are even greater for ground vehicles such as passenger cars and trucks. Although the cost of simulators for these vehicles is not nearly as great as that of aircraft simulators, in most of the circumstances they will exceed the cost of the vehicle itself. This relatively high cost results in a greater reluctance on the part of organizations to invest large amounts of money in driving simulators if the same training can be done in the actual vehicle. Thus, the only real justification for using driving simulators is to allow training for hazardous conditions in which exposing the driver and the real vehicle cannot be justified. This usually means that driving simulators in civilian applications are generally restricted to professional driver training and evaluation. Thus, vehicle simulators are most likely to be found supporting driver training for police, firemen, ambulance drivers, and heavy trucks.

Military ground vehicle applications are less affected by cost than the civilian applications due to the exigencies of large-scale, combat-oriented training that is often required. The use of vehicle simulators in tank and other armored vehicle training, for example, although certainly affected by cost, is more readily justified to save on the high cost of operating these heavy and expensive vehicles.

Passenger car driver training and testing are almost exclusively done within the actual vehicle. The cost of this training is relatively low and it is difficult to justify even a modestly priced driving simulator. This is the case despite the fact that important training in hazard perception and avoidance could be done much more safely and effectively in a driving simulator. Nonetheless, it is relatively rare to find a driving simulator of reasonable sophistication incorporated as a part of normal driver training in the way that simulators are used in pilot training. Although there are many examples of inexpensive driving simulators available, which could provide valuable training and testing, their lack of physical similarity to the actual vehicle may be an impediment to their widespread use. High physical fidelity of a simulator is often associated with its value as a training device. Matching appearance and function of an actual vehicle, however, substantially increases the cost of the simulator, typically by several orders of magnitude. This high cost means that for the average driver, the likelihood is that, outside of gaming devices, they will never use a driving simulator.

Vehicle simulators have long played a role in research, including the design of roadways and airports, vehicle handling qualities, vehicle instrumentation and cockpit design, and in the behavior of vehicle operators under a variety of conditions. The ability to systematically study variables of interest under controlled conditions and to repeat those studies with differing groups of vehicle operators is essential in science and engineering. This ability to control and replicate conditions makes the vehicle simulator a powerful tool in the development of engineering and scientific knowledge.

FIDELITY IN SIMULATION

Throughout the history of vehicle simulator development, the guiding principle in their design has been to achieve the appearance and functioning of the simulated vehicle. This design concept is generally referred to as fidelity. Broadly speaking, fidelity, as the term is normally defined, is faithfulness to the original. Every aspect of the vehicle simulator, therefore, should be faithful to the original if it is to have high fidelity. A passenger car simulator should appear and function as a typical passenger car would appear and function. If the passenger car simulator is defined as a specific make and model then it should appear and behave as if the actual make and model would appear and behave. This design goal applies to all vehicle simulators, including aircraft simulators. In flight simulators, high fidelity would include the actual instrumentation and controls of the real production aircraft as well the same seats and control devices of those aircraft. Cockpit layout and dimensions would be exactly the same as the actual aircraft.

This approach to the design of vehicle simulators with its emphasis on appearance and function has a very long history in simulator design. Indeed, as a design goal, this near obsession with objective or *physical fidelity* has dominated the design of vehicle simulators from the beginning. The emphasis on this design strategy is the major reason why the cost of these devices is so very high. There are a variety of reasons why physical fidelity dominates vehicle simulator design and why this design goal may need to be questioned.

Physical Fidelity

The adoption of what is termed *physical fidelity* in vehicle simulator design is likely a logical outcome of what might be described as a pursuit of physical replication in design. The closer the copy is to the actual object the more successful it is as a simulation of that object. Thus, wherever possible, every aspect of the original should be replicated in the simulated object or system. Only when full physical replication is achieved, it is argued, can the full benefits of the original be achieved. Anything less than full physical replication will result in some, often undetermined and unpredictable, reduction in the value of the simulated system as a viable surrogate for the original.

From a design engineering standpoint, physical replication of the desired system and its environment is a relatively straightforward application of known engineering principles. The physical movements and forces acting on the vehicle are quantifiable as are the dynamics of the vehicle, which determine how the vehicle will respond to these forces. As photorealism of the external visual scene is also a goal of physical fidelity, one need to compare only the resulting simulated visual scene to a photo of its real-world counterpart to determine its fidelity. This is often done as a means of verifying the photorealism of the simulator's visual imaging system.

Physical fidelity as a design goal not only has the advantage of a design analog that uses the real world to define its specifications, it also has the advantage of appealing to an intuitive sense of what the simulated environment should be, that is, as similar as possible to the real world. Within the framework of human experience, it is logical that a simulated environment, physically indistinguishable from the real world would achieve the maximum possible value in training, testing, and research. Thus, physical fidelity not only provides a clearly defined design goal in its use of the real world as an analog but also serves a very attractive marketing and sales function. As most consumers of virtual environment technology view physical fidelity as a measure of realism and as realism drives the value proposition of virtual environments, it follows that physical fidelity is a highly desirable design goal from a purely marketing perspective as well as an engineering one.

Technical Limits and Cost

From the designer's perspective, the degree of physical fidelity can, and often is, defined in terms of the available technology and, more importantly, cost. Eliminating or degrading certain designed components is often arbitrary with little thought given to the consequences of the design change on the behavior of the end user. Instead, the cost of design components is matched against the total cost and those components that are deemed less important for physical fidelity are discarded from the design. For example, driving simulators that are deemed high fidelity will include at least the full vehicle cab with instrumentation and control devices that are the same as the real vehicle. The same approach is used in aircraft and ship simulators where high physical fidelity is desired.

Physical fidelity with its emphasis on physical replication becomes problematic when it encounters the inevitable limitations in simulation technology. Limitations in visual display resolution necessarily result in the loss of image detail. Since loss of this detail results in a less than photorealistic image of a visual scene through the vehicle windshield, physical fidelity will suffer. From the standpoint of design decision-making, there is no means of determining what effect this has on the overall value of the device other than it is somewhat less than high fidelity. The loss of physical fidelity may or may not affect subjective realism of the displayed image and it may or may not affect the behavior of the device user. The consumer of this technology, however, is undoubtedly influenced by the degree of physical fidelity and is more likely to pay more when it is present than when it is not.

As we will see in the succeeding chapters, the emphasis on physical replication as a design goal has often lead to costly and ineffective design components. This has been especially true for motion platform systems which provide only a modicum of subjective realism with little or no effect on user behavior for most vehicles. Even those platforms that produce behavioral changes often do so only with advanced motion platforms systems which are prohibitively expensive.

A similar case applies to the design of visual imaging systems in which photorealism drives the design. Many tasks performed in vehicles simply do not require this level of detail in the visual image. Similarly, the stereopsis produced by the binocular image display is often touted as essential, but analysis of the depth cueing properties of stereoptic displays is negligible beyond 2 m. This makes this design proponent of dubious merit for vehicle simulators though high physical fidelity would make it a requirement.

A notable exception to the obsession with physical fidelity is in the area of force feel and feedback in manual control. Many device specifications do not even address the issue and often those that do, do so in a purely cursory manner. It is likely that this is due to the fact that force feel is often integrated within perceptual motor control to the point where differences in feedback between the real and the simulated are often difficult to discern. It is also possible that this area of design is, despite its importance to vehicle control, not obvious to the purchaser of the device. As a result, control force and force feedback are not ranked high in assessments of simulator physical fidelity. Many less expensive vehicle simulators often sacrifice control force and feedback perceptual fidelity as means of reducing cost.

The emphasis on physical fidelity in the design of vehicle simulators as well as other virtual environments leaves the designer with no real ability to justify design decisions. If the physical fidelity of the device must be reduced because of cost, there is little guidance as to which design component should be sacrificed. It is vitally important that design decisions be based on objectively defined criteria as is the case with any other engineering activity. Of course, the design engineer will be subjected to marketing and sales pressures as with any product. However, it is important that the designer be able to justify design decisions based on the utility value of the device for the use to which it ultimately will be put. At some point in the product life-cycle, the device will be evaluated based on how it affects the behavior of the people when we use it. This fact should be a factor in the design decision.

The emphasis on physical fidelity of vehicle equipment and its operating environment in simulator design does not include any measure of how this fidelity impacts the end user of the device. Any divergence from physical fidelity therefore requires a testing of the device in order to determine what the effect of the elimination of one feature or another would have on the behavior of the device user. In the area of military vehicle simulators such as flight simulators, this has led to a host of studies attempting to determine what the consequences were for pilot training when one design component or another was either eliminated or in someway degraded. As will be seen in the next chapters, these studies often contradict the assumption that high physical fidelity of a vehicle simulator results in a high degree of either training or of performance comparable to that of a real vehicle. Increases in the physical fidelity of the simulated vehicle equipment, the simulated visual environment outside the vehicle, the simulated physical motion, or other aspects of the simulated vehicle does not necessarily have any effect on, for example, the rate at which an operator learns a task or the degree to which the learning transfers to the real vehicle.

The seeming disconnect between physical fidelity and how the user of the device might behave is perhaps not so surprising when it is considered that the measure of physical fidelity is only indirectly related to what the operator uses to operate a vehicle. Physical fidelity measurements are about those things that are physically measurable in both the vehicle equipment and its operating environment. A reduction in the physical fidelity of a vehicle simulator will not necessarily affect user behavior. As physical fidelity is driven by technical and cost factors, the effects on user behavior are just as likely as not to be due to chance as any other factor.

Alternative Measures of Fidelity

A measure of fidelity that is directly relevant to the behavior of the operator rather than one that simply attempts to replicate real-world conditions is needed if the goal of the vehicle simulator design is to produce behavior that is comparable to real-world vehicle operation. Behavior of the vehicle operator is, after all, what is measured during training, testing, and research. If the conditions that support a given behavior in the real vehicle are present in the simulator, the likelihood is high that the same behavior will be produced in the real world.

At least three distinct measures of fidelity are based on behavioral measures rather than physical measures of fidelity. The first of these, sensory fidelity, focuses on designs which will replicate the same or similar sensory functions of the vehicle operator under comparable operational conditions. Sensory fidelity is generally driven by specific models of the human sensory system. These models mathematically recreate the functioning characteristics, limitations, and capabilities of a sensory system and then evaluate how this model will behave when various design parameters of a vehicle simulator are manipulated. For example, a model of the vestibular system will respond in a certain way to specific levels of acceleration. Alterations in the characteristics of a motion platform system for vehicle simulator can be evaluated against this mathematical model to predict how the vestibular system of the vehicle operator will behave. Sensory models of vehicle operators are both complex and numerous. They exist for both driving and piloting functions. For a recent review on driver sensory modeling see Nash et al. (2016).

Measures of sensory fidelity can also be much simpler than that involved in a mathematical model. They may use only a specific parameter with a receptor system such as vision to determine some limit in a design component. The most commonly used example of this is when the term *eye limited* is applied to a visual imagery system display resolution. This means that the display system imagery will meet or exceed the level of detail resolvable by the human eye. This approach to fidelity sets the design limits based on specific limits of human sensory capabilities. The visual scene simulated in this case will only need that level of resolution detail that the user with normal visual acuity will be able to resolve.

The second measure of fidelity goes beyond the level of a sensory system function and includes higher levels of cortical function. Perceptual fidelity involves processes more complex than that revealed in modeling of a sensory receptor system. It defines simulator fidelity in terms of the limitations and capabilities of human perception. The term *perceptual fidelity* is used here to describe the fidelity of a vehicle simulator design in terms of the degree to which it supports those perceptual processes and perceptual cues, which are part of the vehicle operator's tasks. These are processes involved in such tasks as distance judgment, the closure rate of objects, the perception of self-motion, and others that are normal part of vehicle operation. A central tenet of perceptual fidelity of a behavior-based vehicle simulator design is that the design will support the operator's perceptual and perceptual-motor behaviors. Perceptual fidelity is sometimes referred to incorrectly as cognitive fidelity. Although perception may include some element of higher cognitive activity, it is primarily perceptual processes that are at issue.

Cognitive fidelity, the third measure of fidelity, assesses the design of a vehicle simulator based on its ability to support higher levels of human functioning such as decision-making, problem-solving, and navigation. Once again, as with perceptual fidelity, some aspects of cognitive activity may involve the use of perception; however, the primary element or the component of the task is cognitive. Pilots, for example, may rely on a variety of information to decide on diverting to an alternative airport. For cognitive fidelity, the design of the simulator should support this decision-making process by providing the pertinent information. The form in which that information is presented is not essential, but both the availability and accurate timing of the information are necessary to support the diversion of the decision-making process.

Perceptual Fidelity

Among the three behavior-based measures of fidelity, perceptual fidelity is the most relevant to the operation of vehicles. Much of vehicle operation involves fairly primitive perceptual processes, that, although more complex than the functioning of a simple receptor, involve systems that have been subject to investigation for many years. Perception, unlike sensation, is an active process of selectively extracting information from the environment. This activity is essential to the vehicle control process. It is also essential to the safety of the vehicle that the operator is searching the visual scene ahead for potential collision hazards.

Perception is not always a conscious act. Much of perception occurs without conscious attention such as postural control and balance. Furthermore, experienced vehicle operators will automate many skills, particularly low-level perceptual-motor skills. This is a common consequence of overlearning of skill which over time and many repetitions results in the skill been executed without conscious thought. The experienced driver's use of the accelerator and brake pedals is one example and the use of clutch and gear shift is another.

PERCEPTUAL FIDELITY AND DESIGN

It is argued here that perceptual fidelity is superior to other types of fidelity measures of a vehicle simulator in which the goal of that device is to produce behavior, which is comparable to or indistinguishable from, behavior produced under similar situations in the real world. Perceptual fidelity has a direct connection to essential elements of operator behavior and is therefore the most efficient means by which cost-effective simulator designs can be achieved.

One might ask why, if perceptual fidelity is the most desirable goal of vehicle simulator design, rather than physical or other types of fidelity, it is not always the objective of simulator design. The reasons for this are many. The first of these are that the technologies that support simulator development evolved slowly and at different rates throughout the history of vehicle simulation. This meant that any notable shortcomings in the effectiveness of physical fidelity in the areas of training, testing, and research could always be attributed to the immaturity of the simulator technology. Once the technology had achieved maturity, physical replication of the vehicle and its operating environment and therefore, full physical fidelity, would become a reality. This led to substantial investments in research on simulation technologies and comparatively little research into how this ever increasing physical fidelity actually affected behavior.

Second, although physical fidelity has the advantage of a real-world analog definable in physical terms, the behavior-based fidelity metric has no such advantage.

Measures of perceptual processes such as distance perception exist but they do not have the reliability or elegance of a physical equation. Moreover, the relationship between perception and behavior is not necessarily straightforward. In the case of distance perception, the judgment of distance can be affected by a variety of factors both in the scene content as well as in the human visual system itself. To reproduce the behavior of an individual distance perception can, in fact, be done in a variety of ways due to the high degree of redundancy in human distance perception processes. Although this complicates the process of design, it does provide the designer with options in terms of providing cues to produce a specific behavior.

A third disadvantage of perceptual fidelity as a paradigm or model for vehicle simulator design is that very few engineers involved in simulation or in virtual environment design in general are trained in human perception to the degree necessary to use it as a basis for design. A design engineer needs to be as knowledgeable of human sensory and perceptual processes as he or she is of simulation technology. More often, the design team will need to employ a specialist with the requisite knowledge at the beginning of the design process.

In addition, the behavior-based metric of fidelity needs a database of relevant behaviors for that particular vehicle that are obtained either in the real world or in an advanced research simulator. Although data exist on the behavior of drivers and pilots in real-world operations these data lack the level of detail that is necessary to make definitive comparisons between behavior in the simulator and behavior in the real vehicle. This is partly due to the difficulty of measuring behavior in the real-world vehicle and partly because there are very few research simulators that can serve as surrogates for real-world operations. However, with the availability of advanced research simulators, objective performance measures could be collected under a variety of conditions, which could serve as a benchmark for simulator perceptual fidelity.

Despite its disadvantages, perceptual fidelity in vehicle simulator design is much more likely to achieve a cost-effective design in those situations in which the goal of the design is to produce behavior in the device user that is comparable to that of behavior in the real-world vehicle. A question remains, however, as to whether enough is known about human perception and whether that knowledge is sufficiently quantifiable to be useful in design.

Historical Perspective

The scientific study of perception extends as far back as the nineteenth century. Herman von Helmotz, a German physiologist, is credited with initiating the study of sensory physiology and of the founding of the discipline of psychophysics. The latter was the earliest scientific attempt to quantify the relationship between perception and the physical world. In subsequent years, a host of scientific methods were developed to establish lawful, quantifiable relationships between human perception and the physical world.

Measurements of human perception evolved throughout the early history of the science. One was the threshold or the point at which the energy of a physical stimulus is detectable by a human. These threshold values exist for many forms of physical suprathreshold stimuli, including visual (light energy), audio (sound energy), force (weight or pressure), and movement (physical acceleration) among others.

The second important measurement was the quantification of the ability of the human to discriminate among the energy levels of stimuli. The question of how large a change in the level of stimulation is needed to detect a difference provided much more information about the perception than just the threshold measure. This just noticeable difference or JND, similar to the threshold, was assessed for a wide range of stimuli. Research on JNDs for various stimuli continues today.

More important than simple methodology, psychophysics discovered lawful relationships between physical stimuli and the perceptual response to those stimuli, which applied to the full range of stimulus energy. The JND was found to be proportional to the strength of the stimulus used as a standard. Thus, the amount of increase needed for a human to detect a change in stimulus energy was determined to be a constant proportion. This proportional relationship is expressed by the following formula:

$$K = \frac{S_{std} - S_i}{S_{std}}$$

where:

K is a constant

S is the standard

S_i is the stimulus that is to be compared to the standard

The numerator of the equation is the JND. This formula is generally termed the Weber fraction or Weber's law after Ernst Weber, a contemporary of Helmholtz. The Weber fraction has been calculated for different types of stimuli and fits the data reasonably well for many stimuli. The Weber fraction and the JND have found wide applicability in sensory physiology, neuropsychology, and psychology in general.

Psychophysics, with its reliance on sterile laboratory studies of simple stimuli, was challenged by a mid-twentieth century movement, which stressed the importance of ecological validity over strict scientific control. That is, perception should be seen for its importance to the support of species survival not simply as a laboratory exercise. Moreover, the emphasis on isolated laboratory studies of psychophysics, although important for experimental control, could not capture the rich tapestry of real-life perception, which is often multisensory and quite complex.

Two competing views of ecological perception have evolved. One of these was formed around Egon Brunswik. Brunswik's view of how perception works stresses the importance of the probabilistic nature of perception (Brunswik, 1955). For example, a visual image can provide at best only estimates of actual physical distance. These estimates or cues support a judgment in a probabilistic and not deterministic manner. Cues for distance are not direct measures of the distance of an object, for example. They can only contribute to the judgment of distance and this judgment is always probabilistic. Some cues are better predictors of perceptual tasks such as that of distance judgment than others. Cues that the perceiver relies on more heavily in a particular judgment have higher cue validity for that individual.

The strength of the relationship between the cue or proximal stimulus and the perceptual judgment can be measured experimentally by determining the strength of the correlation between the two. The higher the correlation between the cue and the perceptual judgment, the greater the value of the cue will be in supporting the perceptual judgment. Moreover, if there are multiple cues supporting the judgment, the cue that has the strongest relation to the judgment, that is, its relative cue validity, will play a greater role in the judgment than those with lower validity.

However, humans can and do use cues in perception to an extent far greater than they actually deserve. Some cues may, in fact, have poor ecological validity. That is, in the case of distance, they have a poorer predictive validity with respect to the actual physical distance than other distance cues. Cues or proximal stimuli, which have a stronger predictive value of the real distance of an object have a higher ecological validity for the perceiver. If the cues used by the perceiver are given a higher weighting than their ecological validity warrants, the perceiver is behaving in an ecologically nonadaptive manner. When the human perceiver uses cues or proximal stimuli in a manner that has a high degree of correspondence to their ecological validity, the individual is behaving in an adaptive manner. That is, their perceptual judgments will optimize their own survival. Thus, for this view of ecological perception there are always two functions involved. One is the actual use of a cue in making a perceptual judgment and the other is the validity of that cue with respect to its true utility or value in predicting the real, physical phenomena involved.

Following Brunswick some years later is the notable psychologist James J. Gibson (1979) who also adopted the ecological view of perception but declared that the intermediate proximal stimulus was an unnecessary complication. Rather, the world presents itself visually, for example, through the optical array of the visual sensory system. The optical array is the direct light energy from the world. Humans perceive the world by this direct perception of optical invariants. Direct perception, not a probabilistic inference, describes the way in which perception functions. Gibson's study of the optic flow of elements in the optical array which is caused by the flow of perceivable elements across the retina reveals the critical importance of this visual phenomena to both the perception of self-motion and the perception of heading or direction of movement. Gibson stressed the importance of invariants such as optic flow in the optical array as being the critical elements, which support human perception. Gibson's view of perception might be described as deterministic with regard to adaptation. The role of powerful invariants such as optic flow removes the uncertainty that is normally associated with probabilistic functions.

Although the theoretical differences between Brunswick and Gibson are beyond the scope of this book, it is important that both scientists emphasized the role of perception as essential for human adaptation and survival. The human perceiver uses perception in such a way as to maximize their information in adapting to changing real-world conditions. This characterizes perception as having a much more active role than as a simple passive receptor of information as might be inferred from psychophysical studies.

Second, both of these men stressed the importance of studying human perception and human behavior, in general, in the context of the real world rather than isolated in a laboratory. This approach has seen an increasing adoption in vehicle simulation as evidenced by the use of research simulators as devices for the investigation of human perception in driving and flying. The importance of the use of research simulators is not only that they allow more ecologically valid investigations into complex and multisensory environments but also that by doing so, they increase the external validity or generalizability of the findings to the real-world applications. The use of research simulators in the area of scientific research further emphasizes the importance of designing simulators with high perceptual fidelity.

Bayesian Inference and Perception

The development of psychophysics with its emphasis on scientific rigor and quantification combined with an increasing emphasis on ecological validity has been followed by developments in the application of statistical inference to human perception. Within the past two decades increasing evidence has accumulated that the human perceiver is behaving in a statistical optimal fashion, specifically as a Bayesian inference system. As with Brunswik's probabilistic functionalism, the Bayesian view does not view perception as direct with regard to the physical world but as an inferential process.

The world is highly structured. It exhibits regularities, whereas human sensation and perception operate as a noisy receptor system. This is due to the way in which the sensory system is constructed and the many ways in which incoming information can be distorted. The result of these differences means that although the world presents a well-ordered system of information or cues, the human perceiver can only estimate the value of the information in a probabilistic manner. The combination of the two can be likened to radio reception where the world sends a clear signal but the radio receiver adds random noise. The result for the perceiver is that the actual signal can only be an educated guess or estimate, specifically, a Bayesian perceptual estimate.

Statistical inference, of course, is dependent on the variance of the underlying distribution. For Bayesians, perception is a process of sampling from an existing distribution of perceptual estimates or base rate of values corresponding to a given perceptual cue. This base rate distribution or *prior* of a cue is composed of a measure of central tendency (e.g., mean) and a measure of variability (e.g., range or standard deviation). The perceiver weighs the reliability and therefore the utility value of a cue based on the variance of the cue's underlying distribution. The more variability in the cue distribution, the less weight or importance will be assigned to the cue. When more than one cue is present, the perceiver will rely most heavily on making the perceptual judgment on the cue that has the lowest variance and therefore the highest reliability.

The Bayesian view of perception has strong empirical support. Perceptual judgments do, in fact, appear to be influenced by the reliability of the cue. More important, for the purposes here, the Bayesian approach allows for the combination of rigor and quantification to be applied to solve design problems in perceptual fidelity. Through its use of measures such as the JND and other empirical measures of perceptual performance, it makes possible direct quantification of behaviorally based measures of simulator design such as perceptual fidelity. In the subsequent chapters, the Bayesian method of perceptual weighting will be shown to be of significant value in determining the relative weight or contribution of perceptual cues and, by extension, the relative importance of specific vehicle simulator design components.

Robotics and Machine Vision

The rapid developments of the robotics industry and particularly machine vision have renewed interest in how humans perceive the world and move within it. This has led to ever more sophisticated computational models of how human vision works and how those models might be applied to robotics. This research and development activity has already seen benefits in the quantification of perceptual cues to distance such as texture–density and perceptual cues critical to the perception of self-motion, such as optic flow. Quantification of these and other cues may lead to more accurate assessments of the impact that perceptual fidelity has on the users of vehicle simulators. Simulator design engineers should become more acquainted with the field of machine vision and other attempts to replicate human vision computationally.

Validation and Physical Fidelity

If physical measures of a vehicle simulator are fully realized in the device, the job of the design engineer is effectively complete. For the purposes of physical replication, provided the physical replication meets the design specification, further validation is not necessary. As physical replication is the goal, the fact that a human operator is the end user of the device is fundamentally irrelevant to the issue of design validation. Strictly speaking, there is no need for a human operator evaluation nor is there even a need to discuss issues of how the design of the simulator might affect the behavior. The issue of behavior is outside the province of the design engineer because nothing in the original design specification made any reference to human behavior or human limitations and capacities.

The logic of this viewpoint from an engineering perspective is compelling. It avoids the messy issue of having to deal with humans at all. As one flight simulation engineer described them, humans (pilots) are *adaptive, nonlinear, and inconsistent* (Schroeder, 1999, p. 6). Avoiding the problem of the human operator in system design is not, of course, limited to simulators but pervades most of the engineering disciplines. Adopting an approach that legitimizes physical replication as a design goal is a means by which the design problem can be contained within the bounds of existing engineering knowledge and practices.

In reality, complete physical fidelity is rarely, if ever, achieved in vehicle simulation design. Instead, one or more design components fall short of full replication for reasons of technical limits or budget or both. Visual imagery systems are generally not photorealistic and motion platforms designed to simulate *full physical motion* often produce only a fraction of the physical motion that the real vehicle would experience in the real operating environment. The design engineer can, of course, deny responsibility for the consequences of this less than desirable level of physical fidelity. The customer or perhaps a federal regulator or someone else has made the design decision and must live with the results.

The consequences, unfortunately, are particularly problematic for those using simulators for training and evaluation. Much of professional flight training, especially military training, relies heavily on vehicle simulators. As a rule, all these vehicle simulators are subject to operational testing and evaluation (OTE). The OTE evaluates the training effectiveness of the device in the actual operating environment. Thus, a flight simulator designed for training and evaluation of helicopter pilots would be evaluated in a training program, including training aircraft, for these particular pilots. The ability of the device to support the skills training objective and the degree to which skills get transferred to the real aircraft are measured. This validation process measures the rate at which specific tasks are acquired in the simulator as well as whether, and to what extent, this training facilitates skill proficiency in the real aircraft.

The vendor or manufacture of this device can argue that they have met the requirements of the design specification, which was approved by the training organizations. As the manufacturer of the device builds *simulators* and not training devices, they cannot be held responsible for whatever results are obtained in training regimes. High physical fidelity in the simulation of a vehicle and its operating environment may have little or no influence on the rate at which a skill is learned. Indeed, most factors affecting learning rate such as trainee motivation, performance feedback, instructor teaching skills, and instructional technologies have nothing to do with physical fidelity. The designer, of course, has no control over and cannot be held responsible for the competence of the instructors that use the device. Nor can they be held responsible for the motivation of the trainees.

For decades, military training organizations have tested the operational effectiveness of vehicle simulators. Although these devices provided some effective training, they often fell short of what might be expected of a device that often costs millions and sometimes tens of millions of dollars. One of the vehicle simulator design components that cost the most but has been found to be the least effective, and sometimes wholly ineffective, is the motion platform system. Although the motion platform system is a marvel of modern technology (Lee, 2005; Rolfe and Staples, 1988), it has been shown, from a training perspective, as largely wasted money for most of the fixed wing aircraft simulators. The effects of these devices on pilot training for commercial and military fixed wing aircraft are effectively nil (deWinter et al., 2012). Motion platform systems added millions of dollars to the initial and life cycle cost of these devices. Despite the overwhelming evidence of failure to provide any benefit to training, the requirement of motion platform systems for flight simulators remains in effect for commercial airline training. This requirement appears to be largely driven by the need for physical fidelity in these devices, by pilot opinion that they are essential, and by the fear that removing these ineffective components might somehow result in an accident. Certainly, bureaucratic inertia that resists regulatory change once established has played a role as well.

At this point, the question needs to be asked why, if the design specifications clearly reflect higher physical fidelity, as with the provision of physical motion, that no benefit in training is realized from its presence. The answer, as will be seen in the following chapters, is that the vehicle operator is not necessarily responding to a stimulus just because it happens to be present. Rather, a selective extraction of information from the environment is the rule. The basis of that extraction is determined by the task that the operator is performing at the time. Furthermore, the dynamics of the vehicle may also intervene such that a stimulus may seem to play a useful role for one type of vehicle yet may not necessarily do so for another. This has been clearly demonstrated for the case of fixed wing aircraft when compared to rotorcraft (helicopters).

The second validation problem for physical fidelity arises when the need is for a generic device. That is, the design of a device, which is not specific to a particular vehicle, but one belonging to a general class of vehicles. Many small aircraft simulators are generic for a class of smaller aircraft. For example, a generic flight simulator might be built for aircraft which have a single engine and have a gross weight under 6000 lb. The specifics of a particular aircraft's make or model appearance and function are not important. The use of a generic device, such as one used for instrument training, allows for mass production of devices in volume. The resulting production volume significantly reduces the cost for each device. The original Link trainer was a generic device. It did not attempt to replicate a particular aircraft but provided only basic control devices and instrumentation. It had a very primitive motion platform, generally regarded as ineffective, and possessed no visual imagery system. Pilot trainees relied solely on instruments to operate the device. Despite these large departures from physical fidelity, it provided effective and relatively inexpensive instrument training to a large number of pilots.

Generic devices are even more common in the world of driving simulators. The cost of replicating a particular city or roadway is avoided by construction of generic roadways and streets, which provide the essential conditions for specific training scenarios. Driving simulators used in training and evaluation generally do not have motion platform systems. Many also do not have the full vehicle cab one would expect in a vehicle nor do they have the full visual field of view that a real passenger car or truck would have. Despite their low fidelity, such devices have proven valuable for training and have found to be comparable in effectiveness to high fidelity devices (Allen et al., 2003).

This is not to say that there is no relationship between physical fidelity and behavior. Only that the relationship is far from perfect and this imperfection results in unnecessarily high costs of simulator development. Ignoring the behavioral consequences of design in vehicle simulators has proven to be an expensive mistake. The reality is that all vehicle simulators are eventually subject to behavior validation. As such the design process must reflect the importance of supporting the tasks of the vehicle operator behavior in order to develop an effective device.

Validation and Perceptual Fidelity

Behavior-based design paradigms such as perceptual fidelity place the importance of vehicle operator behavior as a priority in design. This does not exclude physical fidelity, per se; it merely places the need to support operator tasks as having priority over physical replication in device design.

As perceptual fidelity task requirements are defined well before the design specifications are written, the process of behavioral validation should not present unanticipated operator performance problems with the device. Supporting the operator's tasks so that a level of performance comparable to what would be expected of an experienced operator can be achieved significantly reduces the likelihood that the device will fail eventual behavioral validation. Those devices designed with high perceptual fidelity are much more likely to succeed in their intended purpose than those that are not.

Devices that are successful despite being low in physical fidelity are so because they provided the operator with the perceptual cues necessary to carry out the required tasks. This is not to say that this will result in a low cost device. It does mean, however, that the cost of the device will result in a net benefit to operator performance rather than a cosmetic embellishment.

A high degree of perceptual fidelity is achieved when the vehicle operator is provided with the perceptual cues necessary to perform all the tasks required for the device. This, it is argued, will result in behavior that is comparable to that expected of an experienced operator under the same conditions in the real world. By an experienced operator, the reference is not to the professional test driver or test pilot, but to the vehicle operator which is representative of the population of intended device users. Thus, behavioral validation of the device is determined by the performance that is typical or representative of the user population and not, as with the test personnel, of unusual skill levels or experience.

Behavioral validation of a device requires objective performance data from a large, representative sample of vehicle operators. This can be collected from real-world operations where possible. The inclusion of modern data recording devices in vehicles makes this relatively easy. The data collected should also include the conditions under which the data were collected such as weather and geographic location, as such conditions might affect the performance of the vehicle operator.

Other sources of performance data include the use of advanced research simulators where available. These simulators provide the advantage of full control over scenario conditions. This makes data collection in research simulators inherently more reliable than field data. They also provide comprehensive data collection capabilities, so that a full range of vehicle operator performance measurements can be recorded. Nonetheless, care needs to be exercised in assessing the quality of the design of these research simulators. A need for a high degree of perceptual fidelity in these devices is imperative if they are to be used as surrogates for the real-world vehicle and operating environment.

Training Transfer

It is argued here that training transfer paradigm is not desirable for validation of the perceptual fidelity of a device. This is particularly true for the true transfer of training validation where trainees complete training in the actual vehicle following training sessions in the vehicle simulator.

Trainees are, by definition, not experienced vehicle operators and therefore will not exhibit behavior which can be said to be representative of the experienced operator population. Trainee behavior tends be highly variable and unstable as compared to experienced personnel. Even if trained to proficiency, which is often not done, trainee skills tend to be much more subject to interference and decay than experienced operators. Moreover, automation of skills that comes with extensive skill repetition over the years is absent in the trainee and this will affect how perceptual cues are interpreted.

Finally, as noted earlier, a variety of factors affect the rate at which trainees acquire skills. The amount of savings in real vehicle training time that a device demonstrates is one such measure of its effectiveness. Yet this measure is often influenced by factors which have nothing to do with simulator fidelity, such as trainee motivation and instructor skill. These and other factors are often difficult to control in the transfer of trainee studies and will affect the results.

SUMMARY

This chapter began with a definition of the different types of virtual environments and of the history and purpose of vehicle simulators. This was followed by a general review of the issue of simulator design with respect to fidelity or faithfulness to the real vehicle. The problem of fidelity in the development of vehicle simulators is discussed beginning with the role of physical fidelity as a design goal. Physical fidelity, which defines the degree to which the simulator will appear and function similar to the real vehicle, was compared and contrasted with perceptual fidelity. Unlike physical fidelity, perceptual fidelity has as its objective the design of simulators which provide the perceptual cues necessary for the vehicle operator to carry out the tasks for which the simulator has been designed. The former, physical fidelity, relies on the physical replication of the real vehicle, whereas perceptual fidelity is focused on the perceptual cues provided by the simulator to support operator tasks. A brief historical overview of perception was provided, including recent developments in Bayesian perception.

REFERENCES

Allen, R.W., Park, G., Cook, M., and Rosenthal, T.J. 2003. Novice training results and experience with PC based simulator. *Proceedings of the Second International Driving Symposium on Human Factors in Driver Assessment, Training and Vehicle Design*, IA: University of Iowa, pp. 165–170.

Brunswik, E. 1955. Representative design and probabilistic theory in a functional psychology. *Psychological Review*, 62, 193–217.

deWinter, J.C.F., Dodou, D., and Mulder, M. 2012. Training effectiveness of whole body flight simulator motion: A comprehensive meta-analysis. *The International Journal of Aviation Psychology*, 22, 122–183.

Gibson, J.J. 1979. *The Ecological Approach To Visual Perception*. New York: Psychology Press.

Lee, A.T. 2005. *Flight Simulation: Virtual Environments in Aviation*. Burlington, VT: Ashgate Publishing.

Nash, C.J., Cole, D.J., and Bigler, R.S. 2016. A review of human sensory dynamics for application to models of driver steering and speed control. *Biological Cybernetics*, 110, 91–116.

Rolfe, J.M. and Staples, K.J. 1988. *Flight Simulation*. Cambridge, UK: Cambridge University Press.

Schroeder, J.A. 1999. Helicopter flight simulation platform motion requirements. NASA TP-1999-208766. Washington, D.C.: National Aeronautics and Space Administration.

2 Vision

INTRODUCTION

Humans, similar to most mammals, depend most heavily on vision to interpret the world around them. Vision plays a major role in many routine human behaviors such as postural control, the perception of speed, object identification and classification, the perception of depth and distance, and many others. It is not surprising, therefore, that the development of virtual environments has depended heavily on the development of visual imaging systems. The earliest attempts at visual imaging that could accurately render images that simulated movement through an environment consisted of camera-model systems in which a television camera is tracked along miniaturized model of an environment. Control inputs to the vehicle would command the camera direction as well as its speed as it tracked across the model. Early computer-generated images followed with relatively crude, monochromatic line drawings on the operator's visual display. Only recently technical developments in computer monitors and graphical processing units have provided what might be called a reasonable facsimile of the real world as it might be seen by the human eye, though still short of the photorealism, which is the putative goal of this technology. Many areas of normal human visual perception as yet cannot be matched by the available technology, whereas other areas are well within the current technical capabilities but the lack of understanding of how to properly use the technology limits their use. The understanding of how the human visual system functions and the limits and capabilities of human visual perception are necessary in order to effectively use the available imaging technology.

THE HUMAN VISUAL SYSTEM

Of all the human senses, the human visual system has been the one system that has received the most attention from the research community. As a result, there is no shortage of information on how the system functions at least at the basic sensory level. Higher level perceptual functions, particularly those that affect the sense of movement are, however, less well understood. Nonetheless, a review of how the visual system works at the sensory and the perceptual level is needed in order to inform the design process as to its limitations and capabilities.

The basic sensory receptor of the visual system is the eye. The function of the eye is to convert light energy into the electrochemical processes of the nervous system, which it does with the aid of some 130 million photoreceptors in the retina. Two main classes of photoreceptors exist: one set has evolved specifically for daylight or photopic vision and the other has evolved for night or scotopic vision. During photopic vision, light energy levels ranging from 10 to 10^8 cd/m^2 stimulate cone receptors in the retina. The receptors are capable of trichromatic (red, green, blue, or RGB) color processing. They convert the light energy and transmit it through the optic nerve. The fovea contains the highest density of cone receptors, some 150,000 per mm^2 to 180,000 mm^2 within 1° to 2° of the foveal axis (Jonas et al., 1992). Cone receptors are the primary receptors for color perception and their high density means they are the primary receptors responsible for photopic visual acuity. Cone density, however, drops dramatically in the periphery of the eye by about 80% at 10°, 85% at 20°, 90% at 40°, and finally reaching its lowest level of 97% reduction at 60° off the foveal axis (Dragoi, 2017). Receptor density of the eye is largely responsible for the dramatic loss of acuity of the visual periphery reaching its lowest density level of cone receptors at 2500 per mm^2.

Scotopic vision light levels range from 10^{-6} to 10$^{-3.5}$ cd/m^2. The lack of light energy means that the cone receptors can no longer function. Instead, rod receptors are engaged to translate the limited light energy for neurotransmission. Rod photo-receptors are distributed at a relatively low density across the visual periphery with a mere 30,000 per mm^2. There are no rod receptors in the fovea. Thus, night vision is only available in the peripheral areas of the visual field. Unlike the cone recep-tor distribution, rod photoreceptor distribution off the foveal axis on the temporal side of the eye reaches a low level of acuity (8% of normal) and stays at that level throughout the visual periphery (Dragoi, 2017). The nasal side of the eye declines to a somewhat poorer 5% of normal acuity at 10° in the periphery. Those who wish to see objects more clearly at night are encouraged to view objects slightly off the foveal axis in which the rod photoreceptors are available. Nonetheless, night vision acuity is still seriously impaired when compared to photopic vision. Color vision is also effectively impaired due to the limitations in rod pigmentation. This limits rod receptors to a peak wavelength of some 500 nm or visual purple. This accounts for the color shift to purple that is often experienced in night vision called the Purkinje effect.

Visual processing of light energy is not only determined by these photorecep-tors but by the optics of the eye. The cornea, the first stage of the optical process is a curved transparent tissue that refracts light as it enters the eye. Just behind the cornea is the iris that is colored blue, green, or brown but serving to reflexively dilate or contract the pupil in order to adjust the amount of light energy entering the eye. Finally and perhaps most importantly, is the lens, which serves to adjust or accommodate light rays entering the eyes, so that they are focused correctly on the retina. The lens is adjusted by two sets of small, ciliary muscles attached to each side. The lens adjustments are reflexive based on the elimination of blur in the image. Normally, the lens does its job and the light energy is correctly imaged on the retina. In about 30% of the population, however, correct accom-modation does not occur. Instead, light is focused in front of the retina result-ing in near-sightedness (myopia) or past the retina resulting in far-sightedness (hyperopia). The reasons vary for these phenomena but designers of virtual envi-ronments, particularly those using head-mounted displays (HMDs) should take note as these particular display systems may create problems for users with poor accommodation.

The photoreceptor signals from the retina are transmitted through the optic nerve, optic tract, optic chiasma, and then on to higher brain levels. For control of eye movements and pupillary reflex control, signals are sent to the midbrain. Transmissions important for visual perception, however, are transmitted to the lateral geniculate nucleus of the thalamus. For much of perception, including vision, the thalamus plays an important role as a kind of sensory control station, integrator, and processor. Most of the information that is vital to visual perception as well as to the basic control of eye functions are located in this part of the brain.

The relative large number and complexities of the visual system components mean that the signal transmission takes somewhat longer for visual than for other sensations. The relative neural transmission time of a sensation may influence what type of sensory information reaches the perceiver and in what order. In this chapter on multimodal perception, the relative contribution of each of the sensory systems will be discussed in more detail.

Each human eye encompasses a field of view (FOV) horizontally of about 120° and vertically of about 100°. As the human visual system is binocular, the total horizontal FOV available is about 180°. The maximum overlap of the two eyes is about 60° horizontally.

Visual Contrast

In addition to the available light, visual discrimination of detail also depends on the contrast that exists between an object and its background or between an object detail and other details surrounding it. Contrast may also be produced between color components such as hue, saturation, and brightness that an object or an object detail might possess. In this case, the contrast ratio is between the color component of the object and that of its background. For most applications, simple luminance contrast ratios can be used. For luminance contrast, the formula is as follows:

$$\frac{L_o - L_b}{L_b}$$

where:
   L_o is the luminance of the object
   L_b is the luminance of the background

This formula applies to the case in which the object background is of uniform luminance. In the case in which the background luminance varies, the average luminance of the background should be used. As a rule, the higher the contrast ratio, the more likely an object will be detected in a visual scene. In visual displays, the maximum contrast ratio is limited by the luminance range available in the display. The minimal contrast for photopic vision when using visual displays is 8:1 (Wang and Chen, 2000).

Visual Acuity

Visual acuity is the capacity of the eye to see detail assuming that sufficient contrast is available. Typical measures of acuity are measured using the Snellen chart or less frequently the Landolt C. Both of these measure the degree to which gaps in an object (e.g., a letter of the alphabet) can be detected at a specific distance from the chart. The visual angle (V_α) of that gap in the eye can be calculated using the following formula:

$$V_\alpha = 2\arctan\left(\frac{S}{2D}\right)$$

where:

S is the gap size
D equals the distance of the gap from the eye

The same formula can be used to measure object angular size at the design eye point of a visual display.

The standard for normal vision is a Snellen 20/20, which converts to an angular subtense of the letter gap to 1 arc min. This, in turn, may be used as a means of determining the effective resolution of a visual display. The angular size of the picture elements (pixels) of a display then would need to subtend no greater than 1 arc min to be considered as eye limited. In contrast, a visual display with individual pixels subtending 2 arc min would be equivalent to Snellen acuity of 20/40. Many visual displays in virtual environments today have resolutions of about 3 arc min/pixel. A 3 arc min/pixel resolution is equivalent to a Snellen acuity of 20/60. Uncorrected vision at this level is generally disqualifying for most driver and pilot licenses.

Stereo Acuity

An additional measure of visual acuity applicable to the case of binocular vision is stereo acuity. Human vision consists of two eyes with an interpupillary distance (IPD) averaging about 6.3 cm in adults. Consequently, each eye will be viewing a slightly different image, that is, there will be binocular disparity. The binocular disparity between the images is resolved or fused into a single image at high levels within the cortex for those with normal vision. However, a study of college-age adults found that about 2.7% of those tested were effectively stereo blind and no fusion takes place as a result (Richards, 1970). These individuals will not be able to experience the depth/distance effects of stereoscopic displays. Sustained viewing of these displays may also result in discomfort for these individuals.

The ability to detect the disparity in depth of these images is known as stereo acuity. About 97% of college-aged individuals can detect depth differences of two objects with as small as 2.3 arc min of disparity and about 80% can detect depth differences as small as 30 arc s of disparity (Coutant and Westheimer, 1993). Stereoacuity should not be confused with *stereopsis*, which is the perceptual experience of depth that results from binocular disparity.

BASIC VISUAL PERCEPTION

Fundamentals of visual perception are, in part, a function of the sensory characteristics of the human eye and partly due to more complex processes that occur in the brain itself. Basic visual perception includes a wide variety of phenomena that have a significant influence on the use of most virtual environments, including vehicle simulators. These include object detection, recognition, and classification as well as object size and distance perception. The perception of depth and distance, which includes a number of both monocular and binocular cues, is critical to many human activities. Perception of self-motion or vection is, in part, a function of visual perception as is the perception of spatial orientation. The perception of object movement and perceptual judgments of object acceleration and velocity need to be considered as well as the phenomena of motion parallax and its influence on distance perception. Color perception is, of course, one important aspect of visual perception.

Color Perception

As noted earlier, photoreceptors are largely, though not entirely, responsible for what we perceive as color. In perception, color is defined by a specific combination of hue, saturation, and brightness. Thus, we can experience the color red as reflected or emitted light of a specific wavelength or hue between 370 and 730 nm (Wandell, 1995). A particular hue (red) will have varying degrees of saturation or *redness* (the amount of whiteness). In turn, the hue and its saturation will have a specific level of brightness (as in *bright red*). Brightness is measured by the degree of reflected light or, in the case of virtual environments, the luminance output of a display. What is determined to be color perceptually is much more of a subjective response than the color defined by the display engineers. In the latter case, the typical digital color image may provide 24-bit *true* color in the RGB, trichromatic combinations. A 24-bit color is the result of combining the 8-bits made available for each red, green, or blue display element. The result is a possibility of some 16.7 million individual colors when combining each of the three colors consisting of 256 ($8 \times 8 \times 8$) possible combinations. Although the numbers are impressive, they do not necessarily reflect true colors in the sense meant by human perception. They are, in fact, just digital combinations of different colors, which are allowed by the digital monitor and supporting processors. In fact, due to variations in manufacturing quality of digital monitors, significant variation in color perception is likely. Thus, a *bright red* in one monitor may be just a *red* in another due to a variety of factors. Moreover, the luminance range output of even modern digital displays is much less than that which the eye is capable of using in photopic vision. A digital monitor with a luminance output of 300 cd/m² is far less than the luminance range of human perception. Although practical for most computer applications, caution should be exercised when using these monitors for applications in which color perception accuracy is required.

For human color perception performance, the question is how well individuals can discriminate one color from another. Of course, a small proportion of the population suffers varying degrees of color deficiency: about 8% males and 0.2% of females. However, those with normal color perception should be able to distinguish at least the three primary colors (red, green, and blue) and their various combinations. Various estimates of color discrimination have been attempted with widely varying results. Gouras (1991) estimated the number of possible colors that humans could discriminate at 200. Others have estimated the number at about 1 million (Kaiser and Boynton, 1996). With regard to just hue discrimination, Wright (1954) estimated the number of discriminable hues at about 150 suggesting that a wavelength as small as 2–3 nm (given a range of 370–730 nm) can be detected. The difficulty of finding a definitive estimate of color discrimination is due to the large number of possible combinations that result from combining a range of potential values of hue, saturation, and brightness. Many thousands of pairs would be needed in order to allow for individual, pair wise comparisons. It is probably safe to assume the number of discriminable colors in the thousands, but unlikely to be the many millions made possible by 24-bit color, *true* color image displays.

Color perception is also known to be affected by the condition under which the color is perceived. Specifically, accurate color perception declines below luminance levels of 300 cd/m^2 (Hood and Finkelstein, 1986). As the luminance levels decline to mesopic or twilight vision, luminance levels of 3 cd/m^2 or less, the eyes become more sensitive to shorter wavelength (violet–blue) and less sensitive to longer wavelength (red–green) levels. Although the designer may have intended the user of the display imagery to see one color, a quite different color perception will occur.

Depth and Distance Perception

Virtual environments are often identified as three-dimensional or 3D because they provide visual cues to depth and distance. In other words, the display can graphically depict images that have depth as well as width and height. These images are more often not 3D in the sense that the image is stereoscopic. Most visual displays in virtual environments are monoscopic, that is, they present the same image to both eyes. Attempts to convert conventional displays through the use of special eye glasses and binocularly disparate projected images have been available for decades. Only with the advent of stereoscopic, head-mounted displays (HMDs) have stereoscopic displays become widely available at relatively low cost.

Despite the compelling nature of stereoscopic displays and the popularity in video gaming and in other entertainment, the binocular disparity that enables the experience of stereopsis is only a minor cue to depth and distance. Instead, a variety of monocular cues to depth and distance play a major role in vehicle operations. Thus, even individuals with a loss of an eye are capable of competent distance perception even though they have lost a substantial part of their binocular visual FOV.

Relative Size

The visual scene that presents itself to the perceiver has a variety of cues to what is known as *egocentric* distance perception. That is, the perception of distance of an object from the viewer. In contrast, the relative perception of the distance between other objects in the scene is termed *allocentric* distance perception.

One of the more powerful cues supporting the distance perception is the relative size of objects. In egocentric perception, a closer object will appear larger than the same object at a greater distance. Thus, the angular subtense of an object viewed in the scene is an indicator of how close or far that object is to the viewer irrespective of any other cues that may be available.

This is a particularly a useful distance cue when the visual scene is in motion as is often the case in a vehicle simulator. Judgments of relative distance between the vehicle operator and objects are constantly changing and hence the relative size of these objects changes as well. Judgment of the rate of closure between a vehicle and an object based on this changing object size contributes to the estimate of the time to collision between the two. The time to collision estimate is essential to vehicle operator performance and is an element in many tasks.

In the design of images for virtual environments, objects that are a part of the scene are being viewed not as real 3D objects but as objects that are projected on a two-dimensional surface. The perceived size, in the absence of any other cues, is determined by the retinal size that exists at the design eye point. In theory, at least, the retinal size of an image in the real world should be the same as that of the same image in the simulated world of the virtual environment. This is the only means by which object size as such has any meaning in virtual environment displays as there are no other meaningful units of size available in the simulated environment. It does not follow, necessarily, that if retinal size is not comparable between the real and the simulated environment, the virtual environment is dysfunctional. Rather, individuals will adapt to their somewhat distorted world, although the effects of this adaptation are not always known. Finally, although the designer has some control over the sizing of objects in the scene, the designer does not directly control how the object is rendered in movement. This aspect of the object is solely a function of the image generator geometry engine. The geometry engine is the software module that carries out the computations, which allow the proper rendering of the object's size at a given distance.

Texture Density

It was noted earlier that the perceived distance of an object is a function of retinal size in the absence of other cues. The other perceptual cue that has a powerful influence on the perception of object size (and therefore of object distance) is the texture density of the scene elements in the object's foreground. Texture density refers to an array of perceptible elements or objects of similar size and shape. The size and density of these objects change systematically with the distance from the observer. It is well known that an object placed in such an array will appear larger to the observer at the same distance than an object of the same size and distance without the presence of a textured field. On account of the size of the object appearing larger due to texturing, it will appear closer as well.

Distance perception can be surprisingly accurate in the presence of a textured field. In a study by Loomis and Knapp (2003), observer's judged the distances of objects placed in a real-world grass field. Judgments of object distance were accurate up to 20 m and declined after that distance. The power of the texture density cue is due to two factors that make up the texture: the angular size of individual elements and the density of these elements. In the ground plane, the size of individual elements within the texture systematically shrinks in perceived size with the increasing distance from the observer. At the same time, the density of elements within the texture increases with the distance from the observer. It is this covariation in perceived size and density with distance that makes texture density such a strong cue to distance.

Texture density is a strong cue to distance perception in the real world but is often missing from scene renderings in vehicle simulators, even in an advanced research simulators. Road surfaces and taxi and runway surfaces are often textured in the real world due to a variety of factors such as the use of aggregate's in the concrete and asphalt and the tire skid markings on runways. Indeed, one of the first uses of advanced texturing in simulators was the addition of tire markings on simulated runway surfaces.

The failure to provide this important distance cue in vehicle simulators has been due to the high impact on processing capacity as well as the limited display resolution of these simulators. Many research simulators have display resolution on the order of 3–4 arc min/pixel, well below the capabilities of the human eye. Whatever the particular reason, texture density is essential if a high degree of perceptual fidelity is desired.

Object Occlusion

In a visual scene where large numbers of objects are available for viewing, the visual occlusion or overlap of one object over another is inevitable. Occlusion is a strong cue to the judgment of the relative distance of one object when compared to another. In an archival study of distance cues to the relative distance of objects, occlusion was the most important cue by a large margin (Cutting and Vishton, 1995). The occlusion cue is not as valuable as other cues when it comes to egocentric distance judgment of objects as it depends on the presence of other objects in the line of sight (LOS). Occlusion provides only the ordinal distance information; an object can occlude another by meters or even by kilometers but the observer will only see that one object is closer than another but not by how much. The occlusion cue is of limited value in distance judgments, though this cue is arguably the most reliable cue to the relative distance of two objects.

Height in the Visual Field

On a level ground surface, objects appear to become increasingly closer to the horizon as they recede in to the distance. The perceptual cue to distance is referred to as the height in the visual field. The height of an object in the visual field functions only when the observer is on level ground and the object's base is seen to touch the ground. Evidence has been provided that the human visual system is using the angle formed by the LOS to the object and the observer LOS to the horizon (Ooi et al., 2001). Figure 2.1 illustrates how the visual field height of an object is derived.

The egocentric distance (D) to an object based on the perceived height of the object in the visual field by an observer can be calculated by the following:

$$D = \frac{H}{\tan A}$$

H is the eye height of the observer and A is the declination angle formed by the eye line to the horizon and the LOS to the object of interest. The LOS to the horizon is known to be perpendicular to eye height and is parallel to the ground plane. When the observer moves the LOS from the horizon to the object of interest on the ground plane, the observer is able to judge the ground distance D to the object based on this declination angle. For a pedestrian of 1.6 m in height, an object viewed at a declination angle, A, of 45° would be perceived as closer than an object at a declination angle of 30°.

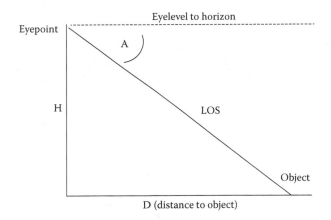

FIGURE 2.1   Distance judgment based on the perceived height of an object in the visual field. Eye height (H) above level ground, angle of declination (A), and distance to object (D).

An automobile driver's eye height is much less than that of a pedestrian. Assuming a driver's eye height of about 1.2 m, the declination angle to the same object as seen by a pedestrian would be significantly less. If the declination angle determines the perceived distance, then a driver would, on average, perceive the distance of the same object as further away than an average pedestrian. Similarly, the driver of a truck with an eye height of 2 m would have a larger declination angle to the same object than either the driver or the pedestrian. Consequently, the truck driver would perceive the same object as closer than either the automobile driver or pedestrian even though objectively the distance has not changed.

The implication of the field height distance cue for vehicle simulator design begins with the simulation of accurate design eye height for the vehicle simulated. Generally, the design eye height is fixed for a particular vehicle's scenic display. This means that those drivers with a different eye height may overestimate or underestimate object distance based on the differences in declination angle. At a minimum, this will reduce perceptual fidelity of the simulator system to the extent that the behavior of an individual driver will deviate from what that driver would experience in the real vehicle. In any case, the effect of differences in perceived distances to due variations in declination angle need to be considered by the designer if perceptual fidelity is to be achieved.

Convergence and Accommodation

As the observer focuses attention on an object in the visual scene, both the eyes will tend to converge on the object beginning at about 10 m and ending at about 10 cm (Owens, 1984). Simultaneously, lens accommodation to bring the object into focus begins at about the same distance and ends at about the same distance as convergence. Both processes are designed to maximize the clarity of the viewed image and both function as automatic processes of the human visual system.

The oculomotor feedback from both the convergent and the divergent actions of the eye and changes in the lens of the eye will send cues from the small, ciliary eye muscles involved. It is believed that these feedback cues may be used by the observer as cues to distance because they would covary in strength with the distance of the object. However, the largest changes occur at about 1 m for both vergence of the eyes and lens accommodation. Both convergence and accommodation appear largely ineffective beyond 2 m with most of the distance cueing limited to about 1 m or less (Cutting and Vishton, 1995). These ranges would be of interest to those wishing to simulate the internal controls of a vehicle cockpit but is not of much use in the simulation of external visual scenes.

Accommodation of the eyes presents particular problems for virtual environments in general and for vehicle simulation specifically. This is largely due to the fact that the simulated display of objects is appearing, not at their actual distance, but at the distance of the display in front of the simulator user. Thus, instead of adjusting the accommodation of the lens to conform to changes to real distances of objects, the lens may only accommodate to the distance of the display from the eye (perhaps 1–2 m away). To avoid this problem, designers have developed infinity optics, such as the collimated lens, so that the eyes will be accommodating to infinity not to the display in front of them. Designers should be aware that, however, this technique has always been controversial and should conduct appropriate tests to assure that the accommodation at optical infinity is in fact occurring when collimation is used.

Linear Perspective

The adage that *there are no straight lines in nature* applies to this well known cue to distance. Linear perspective, the convergence of lines as they recede into the distance, is largely a phenomenon of human culture. Linear perspective can be found in many human structures such as roads, runways, buildings, fence lines, and so on. The distance cue is derived from the interpretation of object size based on its position relative to the point of convergence of the lines. Note that some regard linear perspective as not one cue but a collection of many cues as it often is presented with changes in object size as well as texture density cues. An example of linear perspective can be seen in the converging lines of the texture gradient field shown in Figure 2.2.

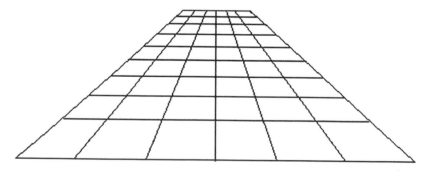

FIGURE 2.2 Illustration of texture in a stylized form. Texture gradients are formed within a visual scene from objects of similar size and shape. The objects become smaller and with greater density per angle of view as the distance from the observer increases.

Aerial Perspective

Objects seen at greater distances are likely to be affected by airborne particulates that scatter light and blur or otherwise partially obscure the object. This interference with light emitted or reflected from the object reduces the object contrast and clarity. As the light scattering effects increase the further the distance of the object is from the observer, the aerial perspective cue is a potential cue to distance. Due to its reliance on atmospheric properties, this is a cue uniquely suited to distances measured in the hundreds or thousands of meters.

Motion Parallax

When the observer is in motion, the differential movement of objects can be a cue to distance. Motion parallax as a distance cue is particularly evident when the observer movement is lateral with respect to the observer's LOS. Figure 2.3a illustrates how motion parallax cues can influence distant perception. Experimental evidence has revealed that motion parallax, by itself, is an important cue to distance. Rogers and Graham (1979) have provided evidence that motion parallax can be as effective as stereopsis (binocular disparity) as a cue to distance.

The display of motion parallax is somewhat problematic for simulators. The typical vehicle simulator displays motion in the visual field in relation to the direction of travel of the vehicle. In other words, the heading of the vehicle determines the nature of the motion parallax. However, true motion parallax changes for an observer as a function of head/eye movements. Thus, only a head-slaved (or eye-slaved) display system can generate veridical motion parallax. Unfortunately, for vehicle simulators using conventional display systems, this is an expensive addition to a distant cue, which may be supplanted by many other distant cues in the visual scene.

RELATIVE EFFECTIVENESS OF DISTANCE CUES

The relative effectiveness of cues to distance has long been an issue because there are such a large number of cues that are available to the observer. As a consequence, there is often a significant amount of redundancy in distance cueing within a typical visual scene. The observer will therefore be faced with several sources in support of distance judgments. Inevitably, the question becomes how effective each of these cues is in supporting the judgment of distance.

The work of Cutting and Vishton (1995) is useful in providing clarity to the issue. These studies focused on the judgment of relative distance of objects in the visual field in ranges of 0.5–5 km. The cues were analyzed for their effectiveness using a variety of archival as well as psychophysical datasets. Datasets included the discriminative utility of the cues in estimates of the relative distance of objects from an observer.

This work revealed that although most distance cues attenuate in effectiveness with increases in the distance to the object, a few cues remain constant in their relative effectiveness. Occlusion, relative size, and relative density maintain constant cue effectiveness over the entire range of 0.5–5 km. Occlusion, although limited to the relationship between the two objects involved, resulted in the highest levels of distance discrimination in JNDs averaging only 0.1%. Relative size discrimination resulted in JNDs averaging 3.0%, whereas the relative density averaged JNDs of 10%.

Using the 10% cut-off level recommended by Cutting and Vishton (1995), we can examine the remaining cues with respect to their range of utility. (A more detailed quantitative analysis of distance cues can be found in Chapter 6). Using this cut-off level, the convergence and accommodation cues are limited to 1 m or less of object distance. This would be roughly equivalent to slightly more than an arm's length. For purposes of vehicle simulation design, these cues are limited in usefulness to judge the distance of objects within the vehicle cab.

Binocular disparity cues are effective up to about 30 m. Motion perspective cues, such as motion parallax, are somewhat more effective than the binocular disparity cues within the first 2 m. Both types of cues then decline linearly in effectiveness up to a distance of 30 m.

Height in the visual field, according to the Cutting and Vishton (1995) analysis, becomes an effective cue at about 3 m distance after which its effectiveness declines gradually in a parabolic function up to 200 m. Throughout its distance, this cue is more effective than either of the three previously described cues.

Although object occlusion dominates as the most potent cue to the relative distance between the two objects, relative size (or angular size) becomes the second most potent cue up to about 30 m.

The aerial perspective cue does not exceed the 10% level until objects are at 500 m or more. This fact suggests that aerial perspective cues are of little value to ground vehicles such as automobiles and trucks in nonmilitary environments. In military environments in which target detection and ranging are important, aerial perspective cues may play a more important role.

The relative effectiveness of distant cues described earlier was evaluated in a single study by Cutting and Vishton (1995) and represents an initial effort to address the relative effectiveness of cues to the distance of objects in close proximity to one another. In addition, the analysis is primarily of static cues viewed by a pedestrian rather than motion cues viewed by a vehicle operator. Despite these limitations, the study is one of the first to present a detailed analysis of the many cues to distance that exist in a typical visual scene.

VISUAL MOTION CUES

In general, the classification of motion cues in human perception falls into two distinct categories. The first category consists of cues associated with an object movement within the scene itself. This is the movement of objects in relation to other objects in the visual scene or allocentric motion. The second category of movement perception is an apparent self-motion or vection. Vection is a visually induced sense of self-motion. That is, it is the illusory perception of self-motion even though no physical motion is present. Clearly, producing vection in vehicle simulators is one of the key elements of perceptual fidelity for tasks that involve vehicle movement.

Object Movement

The detection of object movement in the visual scene is dependent on the displacement of the object of interest. A single object in a group of similar objects can be detected with a displacement of at least 1 arc min within a period of 0.5 s (Watamaniuk and Blake, 2001). Within a structureless field such as a clear and cloudless sky, object motion detection requires about 1.5 arc min of displacement (Legge and Campbell, 1981). Observers can also reliably discriminate an object motion direction in the visual field by about 1°. These findings were found under ideal laboratory conditions in which the observer is stationary. Under these conditions, the ability to detect an object movement is very close to the acuity limits of the eye.

It would not be surprising that the thresholds of an object motion would be different when the observer moves as compared to when the observer is stationary. When the observer moves, as would be the case in a vehicle, the object in the visual scene will move in relation to the observer unless the object is on a direct reciprocal heading. This means that the observer must factor in self-motion when calculating the amplitude and direction of any object movement in the visual scene. As we will see, in real-world vehicle environments, this ability to accurately detect object motion and direction is significantly hampered.

Self-Motion

The second distinct category of motion perception is that of an apparent self-motion or vection. As an observer moves through an environment, the flow of perceptible elements across the retina, or optic flow, creates a sense of self-motion. In the real world, an observer moving through the environment will experience not only the optic flow but also the corresponding sensory stimuli from the vestibular mechanisms of the inner ear, and somatosensory receptors in the feet, muscles, and the joints. Most powerful of all of these cues of self-motion is, however, from the optic flow field.

Different forms of optic flow patterns are shown in Figure 2.3. The precise pattern of the optic flow is formed from the focal point, determined by the LOS of the observer. The flow pattern will radiate outward from the focal point, which is called the focus of expansion (FOE).

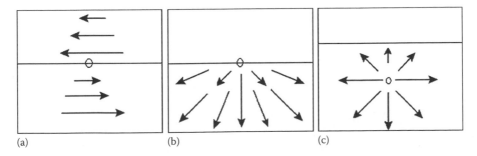

(a)   (b)   (c)

FIGURE 2.3   Optic flow field patterns for (a) translational, (b) longitudinal, and (c) descending longitudinal motion.

Figure 2.3 shows the optic flow pattern for two observations on the ground and one as an aircraft descends to the ground. In Figure 2.3a, the observer is moving from right to left. Note the change of direction of flow in the pattern below the LOS when compared to the flow pattern above the LOS. The typical optic flow on the ground for the longitudinal (in-depth) movement of the observer in Figure 2.3b is what one would see as a pedestrian or operator in a ground vehicle in a forward motion. The flow gradient of objects is always faster when nearest to the observer. Finally, Figure 2.3c shows an example of the flow pattern for an aircraft approaching to land. Note the high velocity of the flow pattern below the LOS due to the forward velocity of the aircraft toward the landing area. The downward pitch of the aircraft attitude eliminates much of the sky above the horizon. The pattern expands upward more slowly above the LOS as the aircraft slowly approaches the ground. This illustrates that optic flow patterns may serve as an altitude change cue as well as a distance cue. Optic flow patterns not only are strong cues that may elicit a sense of self-motion, they can also provide directional information to the observer. It has been shown that observers can discriminate their heading within 0.5° based on the optic flow field (Warren et al., 1988). Extracting heading information from the optic flow field is dependent on aligning the observer's head and eye position, so that the retinal flow is correctly representing the observer movement. This is due to that fact that heading information can actually be disturbed by eye movements, which distort the flow across the retina (Lappe et al., 1999).

VISUAL DISPLAY OF MOTION CUES

The ability to display object motion in a virtual environment is dependent on the rate at which an image is updated. Typically, the update rate is measured by the number of frames or unique redraws per s defined by the number of frames per s (fps). The fps is a term inherited from the filmmaking industry, which set the standard of 24 fps as the frame rate required for motion pictures. In digital imagery, the illusion of motion is produced by turning off and on successive picture elements. The speed of this is partly a function of the processing capabilities as well as the responsiveness of the digital monitor or projection system. The faster objects must move across the screen the faster the update rate will be needed. Thus, the 24 fps standard does not necessarily apply to the digital imagery used in vehicle simulators. To show that a given object motion will be presented smoothly; the formula from Padmos and Milders (1992) can be used:

$$U = \frac{A}{15}$$

where:
   U is the minimum update frequency (in fps)
   A is the object angular speed in arc min/s

The object angular speed is dependent on the object velocity, its distance from the observer, and the angle of the moving object track relative to the observer. Object motion will be greatest when the object motion track is perpendicular to the heading of the vehicle and therefore to the observer's LOS.

Displaying the movement of objects on a reciprocal heading to the observer presents a special problem for vehicle simulator imaging systems. Object motion toward the observer, or *looming* requires an accurate change in object size especially when the looming object speed is high. Display update rates and display resolution need to meet the challenge of accurately displaying looming objects to the vehicle operator in a simulator in order for the operator to accurately assess the rate of closure of an object (such as an oncoming vehicle). If the update rate of the image is too slow or the display resolution is too low, the looming object will appear to jump toward the viewer in discrete steps making it more difficult to judge oncoming speed accurately.

Displaying optic flow is often hindered by the imaging systems restrictions on the number of displayable objects due to limits on graphics processing capacity. Often features seemingly extraneous to the task are eliminated from the visual scene in order to assure that the display update rate remains high enough. This inevitably results in the reduction of optic flow and with it the potential reduction in the sense of self-motion that is so important in achieving perceptual fidelity in vehicle simulators.

It is possible to quantify the amount of optic flow in a visual scene, though the computations are complex. However, the optic flow field changes from moment-to-moment as the elements appearing are occluded or simply disappear and new objects appear instead. Texturing details within objects such as roadways, buildings, and other vehicles add to the optic flow field as do noncultural features such as roadside foliage. Quantifying the optic flow field provides the designer with the means by which rationale trade-offs between the scene content and image processing capacity can be made. Although the computational details are beyond the scope of this book, the reader is encouraged to review the work of Lee (1980) on this topic.

Field of View

Apart from factors such as display resolution and update rate, which might limit the optic flow field, the FOV of the virtual environment visual display will impact the size of the optic flow field available to the observer. For pedestrians, the display FOV should match that normally available to an unrestricted view, around 180° horizontal × 110° vertical. This makes available to the pedestrian the full information available in the optic flow field. It is possible, of course, that the display FOV could be somewhat less than that of the retinal receptor FOV by increasing the number of perceptible elements in the viewable scene, so that a more intense optic flow would be available in the smaller FOV. Ultimately, of course, the effectiveness of the optic flow field on behavior, such as the ability to control the speed of movement, must be measured.

Unlike pedestrian movement in which the FOV is normally unrestricted, vehicle cabs impose substantial restrictions on both the vertical and horizontal FOV of the operator. This means that the operators of a vehicle will have less information from the optic flow field than a pedestrian would have when viewing an identical visual scene. Although unlike pedestrians, vehicle operators have instrumentation such as speedometers that they can consult, the vast majority of operator time in ground vehicles will be spent viewing the visual scene that appears through their windscreen. The restricted FOV of vehicles and, of course, vehicle simulators means that the available information from the visual scene to support tasks such as speed control must be done with greater efficiency and focus than would be the case for a pedestrian.

Driving Simulator Visual Imagery

Perceptual fidelity of the visual scene in driving simulators is measured by the degree to which the visual cues in the simulator compare to those visual cues available in the comparable real-world conditions. If perceptual fidelity is high, then the vehicle operator's performance of perceptual tasks will be comparable in both the simulator and the real vehicle. A driving simulator that achieves high perceptual fidelity will provide all the necessary visual cues to support the tasks required of the particular simulated vehicle.

Object Detection and Image Detail

Drivers operate in a visual object detection environment which typically extends from about 3 m to as much as 500 m in front of the driver. Although the driver needs to be aware of the hazards to either side and to the rear, it is the objects in front of the driver's vehicle that demand the highest levels of display resolution. The 500 m distance is determined by how far away a potentially hazardous object, such as an oncoming vehicle, needs to be displayed to provide the driver with the time to perceive the hazard and to take appropriate measures to avoid it. At 500 m, an opposing vehicle with the equivalent speed as the driver's vehicle at 120 km/h would give the driver of about 7.5 s to decide and execute a decision. An oncoming automobile at this range would subtend a visual angle of about 14 arc min. A smaller vehicle, such as a motorcycle, would subtend about 4 arc min at 500 m. A display resolution of 4 arc min would be sufficient to support the detection of an opposing vehicle, even a small one, to allow enough time for a decision such as a passing maneuver to be made and executed.

Road signage imposes a greater emphasis on the detailed reading and letter recognition. Individual countries typically have standards for letter size and composition, so that road signs can be read at specific distances by the typical driver. For example, the height of speed limit signs in the United States is generally 0.77 m (30 in.). The sign guidance issued by the U.S. Department of transportation is that a sign will be legible at 130 m (40 ft) for every 2.54 cm (1 in.) in height. A sign height of 0.77 m would mean that the sign would be legible at a distance of 369 m (1200 ft). However, a speed limit sign is not legible at that distance, though the sign itself can be seen. To apply this standard to set the minimum display resolution would present unreasonable demands that would needlessly inflate costs with no real benefit. Rather than relying on these guidelines, measures that are more realistic with regard to the actual driver behavior should be used.

A better, more useful measure of signage readability is the same as used in the Snellen eye chart. It is the gaps in the letters themselves that, if discernible, make the letters legible. A rule of thumb is to use 1.5 times the stem size of the letters, which is the width of the letter's vertical and horizontal appendages, to indicate gap size. The capital letter *E* in a sans serif Gothic media font will meet this requirement. Converting these into angular subtense will inform the designer with regard to the distance at which the sign would be legible.

The typical speed limit sign in the United States contains the words *speed limit* and the speed limit itself, for example, below *50*. The stem size of a typical speed limit sign speed number is 4.0 cm (1.6 in.), whereas the information above it has a stem size of 2.0 cm (0.8 in.). Using the 1.5 times stem width, with 1 arc min of resolution this *50* lettering should be legible at a simulated distance of 208 m (675 ft). With 3 arc min resolution, the sign should be legible at a simulated distance of 69 m (225 ft).

An even greater challenge for simulator designers is the legibility of street name signs. The typical street name sign height is 15.2 cm (6 in.) with 10 cm (4 in.) high lettering. The sign guidance would mean that the street name sign would be legible at 73.2 m (240 ft). However, the typical stem size of the lettering is 1.7 cm (0.67 in.). Using the 1.5 times stem width, at 1 arc min resolution, these signs are legible at a distance of 88.3 m (287 ft) and at a distance of 29.4 m (95.7 ft) with 3 arc min resolution.

Understandably, there may be reluctance to expend the extra cost for the high resolution of a display just to allow users to read signs at an appropriate distance. Many simulator designers might simply make the sign (object) larger and with it the lettering it contains. Although this is an understandable work around the problem of poor display resolution, it violates the scale integrity of the simulated visual scene. An increase by a factor of two or three times means that a significant number of objects in the scene are no longer scaled appropriately. These objects will be seen earlier than they should be due to the larger size and will violate the visual scale integrity required for a virtual environment that depends on it for perceptual fidelity. Leaving sign height veridical and altering letter size is a better design alternative.

Object detection and object detail are less of a problem for most conventional applications in ground vehicle simulation. However, certain specialized military applications may require higher levels of resolution for target detection of classification. Despite the evidence of near eye-limited detection of object movement noted earlier, a driver's ability to detect movement of individual objects is much less. Due to unconstrained head movements while driving, detection of an object movement is nearly 3.5 times poorer in real-world driving than it is in the laboratory (Probst et al., 1987). This would place object displacement detection at about 3.5–5.3 arc min for the typical driver. Excluding issues of signage, most object recognition and object detail issues in vehicle simulators can be met with display resolution on the order of 3–4 arc min at the design eye point without significant impairment to tasks such as passing and vehicle following.

The ability to resolve details within a visual scene goes beyond simple object recognition but also includes more subtle cues. Resolution of image details is particularly important in distance and speed control behaviors. Two cues supporting these tasks, optic flow and motion parallax, are often difficult to quantify and require actual behavioral studies to assess variations in image detail and their effects on driver performance.

Optic Flow

One of the behavioral studies which has attempted to evaluate the influence of optic flow in image resolution was conducted by Jamson (2001). In this study, the FOV and image resolution were systematically varied and the effects on driver speed control and lane position were measured. Most importantly, the study used data from a test track to compare to the data collected in the simulator. A double S-curved roadway was used in the simulator and the test track. The use of a curved roadway as a test venue placed more emphasis on driver performance than would be the case if a straight roadway were used.

The study examined three FOV horizontal dimensions (50°, 120°, and 230°) and two levels of display resolution (2.6 and 3.6 arc min). Factorial combinations of FOV and display resolution allowed an assessment of the influence of higher resolution to be examined at each FOV level.

The combination of FOV of 120° and display resolution of 3.6 arc min resulted in driver speed control and lane position closest to that of driver behavior in the real-world test track. Although driving behavior improved when FOV was increased from 50° to 120°, no significant changes were found when the FOV was increased still further to 230°.

The lack of any effect of increasing the resolution of the display from 3.6 to 2.6 arc min is somewhat surprising given that the high resolution should have an increased optic flow by making more scene elements visible. It is possible that object level details were such that improvements in resolution would not increase their visibility. It is also possible that the higher display resolution is not needed given the low level of human visual acuity at the visual periphery in which much of the impact of the optic flow occurs.

It should be recalled that the data were collected on curved roads where LOS of the driver is on the tangent of the curve during the maneuver and is updated continually as the curve is driven (Land and Horwood, 1995). The driver's LOS to the edge of the road would give the driver an advantage when driving with a wider (120°) FOV.

Increasing the display FOV is certainly one mean of increasing the availability of optic flow to the driver provided, of course, that there are elements within the optic flow that will contribute to the flow field. Other investigators have used a different technique to address the flow field issue. This technique keeps the existing FOV but increases the available optic flow field.

The geometric FOV (GFOV) is the FOV which exists within the software and which governs what is actually displayed on the monitor or projection system. This GFOV is theoretically independent of the actual imaging system display FOV in that it can be increased or decreased in size through software changes alone. The ratio of the GFOV to the FOV is generally kept at unity or 1:1 so that the GFOV specified in the software will be correctly displayed on the actual display FOV. Thus, a virtual FOV display of 60° horizontal will be displayed correctly on a display FOV that provides a 60° FOV at the design eye point.

It is possible to change the GFOV: FOV ratio by alterations in the simulator software. If the ratio of GFOV: FOV is smaller, for example, 1.3:1 then the GFOV is minified. That is, more of the available virtual field in the simulator software will be seen on a display having the same FOV. Minifying has the effect of revealing more of the visual scene's optic flow field without an increase in display cost.

Studies by both Diels and Parkes (2010) and by Mourant and colleagues (2007) have experimentally manipulated the GFOV: FOV ratio in order to address a common problem in driver simulators. This is the problem of driver underestimation of speed due, presumably, to an inadequate optic flow field. In both the studies, driver speed control was improved when GFOV: FOV ratio was increased allowing for more of the optic flow field on to the screen. Mourant et al. found a GFOV: FOV ratio of 1.22:1 or a minification of 22% to be optimal in improving the driver speed control.

Generally, the expedient of increasing the GFOV:FOV ratio beyond unity is, however, not recommended. Although the minification of the GFOV may improve the perception of vehicle speed, it does so by compressing the size of objects in the visual scene. Although this does not affect the scaling integrity as such, the rendered object size will be smaller than intended. If the designer's intent is to match the displayed object size to the retinal size, so that objects appear as they would under real-world conditions, then minification will, in fact, make the objects appear to be further away than they should.

Driver Eye Height

Optic flow field effects can be altered by increasing or decreasing driver eye height. As a rule, driver eye height is fixed in the simulator software at a specific point above the simulated surface. The driver eye height is normally that of an average driver eye height for the particular vehicle being simulated.

Variations in driver eye height might be expected to have an effect on the perception of self-motion due to an alteration in the intensity of the optic field flow. The higher above the surface the retina is, the slower the velocity of the flow field will be. In a study of simulated truck driving, increases in simulated driver eye height affected both the distance judgment in a vehicle following task as well as speed control (Panerai et al., 2001). With an increased eye height, vehicle following performance declined along with a tendency to underestimate speed. The study demonstrated that both distance cues associated with surface texturing and optic flow will be affected by increasing driver eye height.

Designers should not arbitrarily increase or decrease the driver eye height in order to alter the distance or speed control cues. Altering the eye height in order to increase the effects of optic flow will introduce undesirable scaling effects. The relationship between the driver and the objects in the surrounding environment is influenced by the driver's perceived height above the ground. When the driver is adjacent to other objects with a known height relationship to the driver, such as a stop sign or a bus stop, the driver expects to see that relationship as it would occur in real life. Arbitrary changes to the driver eye height violate these expected relationships and, in effect, introduce undesirable perceptual artifacts into the simulation.

Motion Parallax

It has been relatively common when discussing driving simulator visual imagery systems to note that these systems do not provide natural motion parallax cues. This is due to the fact that object motion in driving simulators is based on the direction of movement of the vehicle being simulated. Natural or veridical motion parallax of objects for pedestrians depends on the position of the head/eyes while viewing the moving scene.

A study conducted using a driving simulator capable of providing head-slaved motion parallax was conducted by Hultgren et al. (2012). The study assessed driver performance in the passing maneuver with the presence of the oncoming vehicles at varying distances. The choice of a safe oncoming vehicle distance, lateral lane position, and speed control were compared for head-slaved and vehicle-slaved motion parallax. No differences were found in any of these measures for the two types of motion parallax.

Palmqvist (2013) also investigated the role of motion parallax using a research simulator that generated motion parallax based on the position of the driver's head rather than on the vehicle heading. The head-slaved motion parallax effect was evaluated in a passing maneuver. When compared to conventional vehicle-slaved motion parallax, head-slaved motion parallax allowed the driver to move closer to the outside of the lane during the passing maneuver. Although the effect size was very small ($R^2 = 0.06$), the study results provide some support for the contention that the presence of true motion parallax can be beneficial to simulator perceptual fidelity.

Further research is needed to determine whether head-slaved motion parallax has a reliable effect on simulated driving behavior. Moreover, the effect size, if another reliable one is found, would have to be large enough to justify the additional costs associated with the introduction of head or eye-slaved visual imaging systems in driving simulators.

AIRCRAFT SIMULATOR VISUAL IMAGERY

Visual imaging systems in aircraft simulators present a particular challenge because of the large variety of aircraft types and missions. Many of the topics discussed for driving simulators apply to aircraft when operating on the ground. Thus, issues such as speed control, position control, curve and intersection maneuvering, and others involve the same perceptual processes in aircraft piloting as they do in driving.

However, there are clear differences in what is needed in terms of perceptual fidelity when the aircraft pilot is engaged in tasks such as takeoff, airborne maneuvering, and landing. As with driving, there are tasks associated with military operations that also involve perceptual issues that may require specialized visual imaging systems.

Object Detection and Object Detail

Many of the issues that affect the perceptual fidelity of driving simulators also affect aircraft simulators. Object detection is particularly important when it involves the detection of another aircraft flying in the vicinity of the pilot's own aircraft. This has a particular impact on the resolution of aircraft simulator displays due to the greater distances involved.

When visibility is good, pilots are required to *see and avoid* other aircraft even if they are under air traffic control. A small, general aviation aircraft on a reciprocal heading with the pilot's own aircraft will subtend about 1 arc min at a distance of 5 km (3 statute miles). Even then the detection could only occur under ideal atmospheric conditions, which effectively eliminates atmospheric interference. At this distance, the pilot would have 45 s to decide on the avoidance maneuver assuming both are flying at 193 kph (120 mph). The time-to-collision declines rapidly with higher speed aircraft. Detection of larger aircraft would occur at a greater distance with 1 arc min resolution but these large aircraft are flying at higher speeds. An aircraft with a 3 m cross section would be detected at 10 km (6 statute miles) but with each aircraft traveling at 402 kph (250 mph) the time-to-collision would be nearly the same (about 43 s) as it was in the earlier example with small aircraft.

Airport signage details, as with road signage, are regulated by various transportation authorities. In civil aviation, the International Civil Aviation Organization sets many standards as do local national authorities such as the federal aviation administration (FAA) in the United States. As aviation is an international as well as national transportation system, many aspects of airport markings and signage guidance apply worldwide.

The smallest airport signage that a pilot will need to read is a panel 46 cm (18 in.) in height with a legend of 30 cm (12 in.) in height. The stem size of the lettering is about 3.5 cm (1.4 in.). At 1 arc min resolution using the 1.5 rule, the sign is legible at 180 m (585 ft). At 3 arc min, the sign is legible at about 60 m (195 ft). Such a sign might be placed at the side of an airport taxiway close enough to the ground to avoid contact with the wings of aircraft. At a typical taxi speed of 20 kts (37.5 kph or 10 m/s), the pilot will have 6 s to read and respond to the taxiway instruction at 1 arc min resolution but only about 2 s if the resolution is 3 arc min.

Display resolution plays a much greater role in aircraft simulation than in driving simulation largely due to the higher speeds and the type of tasks that are performed. Aircraft used in military operations are likely to increase the requirements for display resolution because of the type of missions flown and the more flight critical tasks that pilots are required to perform. For example, military mission may require not only an early object detection at the furthest possible range but also recognition and classification of an object as a threat.

A number of recent attempts to evaluate the effects of display resolution have been hampered by technical issues. One of these issues relates to changes in contrast and other issues of image quality that may be introduced when display resolution is increased (Geri and Winterbottom, 2005). The second problem is the difficulty of increasing resolution over the wide FOVs typically found in aircraft simulation. As the pixel size is reduced with an increasing spatial resolution, the number of pixels needed to achieve the same FOV increases as well. This has substantial effects on processing resources which will affect image frame rate and other factors.

Field of View

The issues of visual imaging system for aircraft have been investigated for much longer than those for driving simulators simply because aircraft simulator development has been around much longer. A review by Wolpert (1990) of some of these investigations into display FOV showed that, not surprisingly, a wider FOV generally results in an improved pilot performance. For aircraft in general, the studies found marked improvement in tasks such as roll attitude control when the horizontal FOV was increased from 48° to 300°. Roll control is essential to maintain correct bank attitudes in turns, for example. When pilots are flying without the aid of instruments, which aid in attitude control, the pilot relies solely on the external, visible horizon for roll and pitch attitude control.

For lateral target tracking in helicopter hover, Wolpert (1990) found studies showing that a wide FOV of 105° produced better performance than a narrow (10°) FOV. This was attributed to the presence of background feedback from the former, which was largely absent from the latter. In a study of pilot performance in a large commercial airline transport, wide FOV (114°) provided overall better performance in controlling turns and aircraft vertical velocity than a narrow (40°) FOV. Moreover, when the visual scene complexity was increased the wider FOV produced better pilot performance in a turn than a comparable scene with a narrow FOV.

One of the more undesirable impacts of reduced FOV has been on the manner by which tasks are executed. In a field study of FOV in real aircraft operations, both horizontal and vertical FOVs were varied (Covelli et al., 2010). Significant performance effects were seen in runway alignment and vertical tracking when FOV was reduced below 120° horizontal by 81° vertical. Head movements increased dramatically when FOV was reduced even further to below 80° horizontal by 54° vertical. Moreover, pilots spent increasing amounts of time inside the aircraft scanning instruments as an adaptive response to the loss of information from the external visual scene. Adaptive behaviors resulting from FOV reductions have been found in other studies as well. A simulator study of low-level flight performance in a military transport aircraft found that reduced FOV increased the use of aircraft flight instruments and altered their use of information from the external visual scene (Dixon et al., 1990). These adaptive behaviors are strong indicators of how poor perceptual fidelity in simulator design can adversely affect its behavior.

Texturing

One of the dominant shortcomings of early flight simulators was their inability to display high levels of detail such as surface texturing. The runway was represented only as a set of outlines with possibly a centerline and perhaps the runway number. The lack of any other details, especially the lack of texturing of the runway, resulted in poor pilot-landing performance. The texturing that is normally provided by the runways is a combination of factors, including tire markings made during landing and braking, aggregate's used in the runway surface material, and other elements. These texturing details provide depth and distance cues for the critical landing task.

Perhaps the most difficult task for any pilot, certainly for a novice pilot, is the landing of an aircraft. The visual approach requires the maintenance of correct airspeed, alignment with the runway, and the correct glide slope. As the aircraft descends to the runway, the next critical task is to check the rate of descent. The task called the *flare* must be executed by the pilot at the correct time. If the pilot fails to check the rate of descent, the aircraft will slam into the runway with potential catastrophic results. Checking the rate of descent too early, however, will result in the aircraft floating dangerously down the runway. The pilots timing of the flare is therefore critical.

The timing of the flare appears to depend on two critical perceptions. The first is the pilot's perception of the height of the aircraft above the runway. The second is the perception of the time for the aircraft to contact the runway (or time to collision). The perception of height or altitude above the runway is believed to be determined by the pilot's use of the visual angle formed by the edges of the runway to wither the which would be perceived as increasing as the altitude above the runway is decreased.

In the study conducted by Mulder et al. (2000), performance in timing the flare with only the runway outline was evaluated. Mulder et al. noted that the use of the runway outline alone for altitude judgment could be hazardous as runway width varies considerably from one airport to another. They posited that the pilots must be using some other cue to aid in timing the flare. That other cue was determined to be the changes in the optic flow pattern, which occurred as a result of changes in altitude above the runway. The optic flow pattern results from the movement of texture cues from the runway itself. Therefore, the addition of texturing cues to the runway should improve pilot timing of the flare maneuver. The improvement in pilot performance of the flare maneuver with the addition of texture to the runway was indeed confirmed by Mulder and colleagues. The authors suggested that both runway outline and surface texturing are essential in the simulation of runways.

A more recent study found similar results (Palmisano et al., 2006). This study also found that adding additional texturing along the aim line (typically the centerline of the runway) will improve flare performance even more. The provision of texturing to runways in modern simulators has become relatively common due to improvements in computer and display technologies. It is not, however, clear that the amount of optic flow produced by this texturing is sufficient given the relatively low level of display resolution and luminance of the displays used in these simulators.

The provision of an optic flow pattern as an aid to landing performance in simulators is an example of what might be described as a critical cue to task performance. The absence of such a cue in a flight simulator would mean that the pilot's performance would be affected substantially. Moreover, if the pilot was a trainee, the development of an excessive reliance on a potentially unreliable and even dangerous visual cue could result. Thus, the rendering of accurate optic flow fields should be deemed crucial in flight simulator design.

Unlike road surfaces, the runway area is a relatively barren surface excepting the runway centerline, occasional runway lighting, and approach aids. This means that the simulator has to present those texturing details that do exist at a sufficiently high level of perceptual fidelity to support this critical landing task. The advent of new display technology such as ultra high definition (UHD) monitors now available provides the possibility of determining more precisely the level of texturing detail in real-world runways that needs to be visually simulated. These UHD monitors can provide eye-limited resolution (1 arc min/pixel) over a relatively large FOV.

SUMMARY

The human visual system is the most dominant and the most important sensory system that vehicle operators possess. This chapter reviewed the basics of human vision as well as the many variables that affect human perception. The processes involved in object detection, and object perception as well as those involved in the perception of distance are discussed. A detailed discussion is provided on the role of monocular and binocular cues to distance, including the definition of a large variety of cues that typically support distance perception. Perceptual cues that support self-motion such as optic flow and the role of motion parallax in distance perception are described and how these cues are affected by simulator design variables such as FOV and resolution is also described. Visual system imagery of vehicle simulators and how their design impacts the visual perception of the vehicle operator are reviewed.

REFERENCES

Coutant, B.E. and Westheimer, G. 1993. Population distribution of stereoscopic ability. *Ophthalmic and Physiological Optics*, 13, 3–17.

Covelli, J.M., Rolland, J.P., Proctor, M., Kincaid, J.P., and Hancock, P.A. 2010. Field of view effects on pilot performance in flight. *The International Journal of Aviation Psychology*, 20, 197–219.

Cutting, J.E. and Vishton, P.M. 1995. Perceiving layout and knowing distances and knowing distances: The integration, relative potency, and contextual use of different information about depth. In W. Epstein and S. Rogers (Eds.), *Handbook of Perception and Cognition: Vol. 5. Perception of Space and Motion* (pp. 69–117). San Diego, CA: Academic Press.

Diels, C. and Parkes, A.M. 2010. Geometric field of view manipulations affect perceived speed in driving simulators. *Advances in Transportation Studies*, 22, 53–64.

Dixon, K.W., Martin, E.L., Rojas, V.A., and Hubbard, D.C. 1990. Field of view assessment of low-level flight and airdrop in the C-130 Weapons System Trainer (WST). *USAFHRL Technical Report, January 89-9 20*.

Dragoi, V. 2017. *Visual Processing: Eye and Retina.* Houston, TX: University of Texas. Retrieved from neuroscience.uth.tmc.edu (accessed January 18, 2017).

Geri, G.A. and Winterbottom, M.D. 2005. Effect of display resolution and antialiasing on the discrimination of simulated aircraft orientation. *Displays*, 26, 159–169.

Gouras, P. 1991. Color vision. *Principles of Neural Science*, 3, 467–479.

Jamson, H. 2001. Image characteristics and their effects on driving simulator validity. *Proceedings of the First International Driving Assessment, Training, and Vehicle Design.* IA: University of Iowa.

Hood, D.C. and Finkelstein, M.A. 1986. Sensitivity to light. In K.R. Boff, L. Kaufmann, and J.P. Thomas (Eds.), *Handbook of Perception and Human Performance: Vol. 1: Sensory Processes and Perception.* New York: John Wiley & Sons.

Hultgren, J.A., Blissing, B., and Jansson, J. 2012. Effects of motion parallax in driving simulators. *Driving Simulation Conference*, Paris, France, September 6–7.

Jonas, J.B., Schneider, U., and Naumann, G.O.H. 1992. Count and density of human retinal receptors. *Graefe's Archive for Clinical and Experimental Opthalmology*, 230, 505–510.

Kaiser, P. and Boynton, R. 1996. *Color Vision.* Washington, DC: Optical Society of America.

Land, M. and Horwood, J. 1995. Which parts of the road guide steering? *Nature*, 377, 339–340.

Lappe, M., Bremmer, F., and van den Berg, A.V. 1999. Perception of self-motion from visual flow. *Trends in Cognitive Sciences*, 3, 330–336.

Lee, D.N. 1980. The optic flow field: The foundation of vision. *Philosophical Transactions of the Royal Society of London*, 290, 169–178.

Legge, G.E. and Campbell, F.W. 1981. Displacement detection in human vision. *Vision Research*, 21, 205–213.

Loomis, J.M. and Knapp, J.M. 2003. Visual perception of egocentric distance in real and virtual environments. In L.J. Hettinger and M. Hass (Eds.), *Virtual and Adaptive Environments.* Mahwah, NJ: Lawrence Erlbaum Associates.

Mourant, R.R., Ahmad, N., Jaeger, B.K., and Lin, Y. 2007. Optic flow and geometric field of view in a driving simulator display. *Displays*, 28, 145–149.

Mulder, M., Pleisant, J.-M., van der Varrt, H., and van Wieringen, P. 2000. The effects of pictorial detail on the timing of the landing flare: Results of a visual simulation experiment. *The International Journal of Aviation Psychology*, 10, 291–315.

Ooi, T.L., Wu, B., and He, Z.J. 2001. Distance determined by the angular declination below the horizon. *Nature*, 414, 197–200.

Owens, D.A. 1984. The resting state of the eyes: Our ability to see under adverse conditions depends on the involuntary focus of our eyes at rest. *American Scientist*, 72, 378–387.

Padmos, P. and Milders, M.V. 1992. Quality criteria for simulator images: A literature review. *Human Factors*, 34, 727–748.

Palmisano, S., Favelle, S., Wadwell, R., and Sachtler, B. 2006. Investigation of visual flight cues for timing the initiation of the landing flare. *ATSB Research and Analysis Report Aviation Safety Research Grant, Report No. B2005/0119*. Canberra, Australia: Australian Transport Safety Bureau.

Palmqvist, L. 2013. Depth perception in driving simulators. Bachelors thesis, University of Umea, Umea, Sweden.

Panerai, F., Droulez, J., and Kelada, J.M. 2001. Speed and safety distance control in truck driving: Comparison of simulation and real-world environment. *Proceedings of the Driving Simulation Conference, DSC 2000*, Paris, France.

Probst, T., Krafyck, S., and Brandt, T. 1987. Self-motion perception: Applied aspects for vehicle guidance. *Ophthalmic and Physiological Optics*, 7, 309–314.

Richards, W. 1970. Stereopsis and stereoblindness. *Experimental Brain Research*, 10, 380–388.

Rogers, B. and Graham, M. 1979. Motion parallax as an independent cue for depth perception. *Perception*, 8, 125–134.

Wandell, B.A. 1995. *Foundations of Vision*. Sunderland, MA: Sinauer Associates.

Wang, A.-H. and Chen, M.-T. 2000. Effects of polarity and luminance contrast on visual performance and VDT display quality. *International Journal of Industrial Ergonomics*, 25, 415–421.

Warren, W.H., Morris, M.W., and Kalish, M. 1988. Perception of translational heading from optic flow. *Journal of Experimental Psychology: Human Perception and Performance*, 14, 646–660.

Watamaniuk, S.N.J. and Blake, R. 2001. Motion perception. In S. Yantis (Ed.), *Stevens Handbook of Experimental Psychology* (3rd ed.). New York: John Wiley & Sons.

Wolpert, L. 1990. Field of view information for self-motion perception. In R. Warren and A.H. Wertheim (Eds.), *Perception and Control of Self-Motion* (pp. 121–123). New York: Psychology Press.

Wright, W.D. 1954. Defective color vision. *British Medical Bulletin*, 9, 36–40.

3 Physical Motion

INTRODUCTION

The one seemingly essential attribute of a vehicle simulator is that the simulator should, as with the real-life vehicle, physically move. Unlike the visually induced illusion of motion that can be provided by the computer generation of an external visual scene, the addition of physical (nonvisual) motion of the device would seem more desirable. Indeed, in the simulation of vehicles, no issue has been more contentious than whether or not physical motion of the device is, at least at some level, a fundamental design requirement. However, in order to meet the requirement of perceptual fidelity as defined here, the presence of physical motion must affect the vehicle operator's behavior in some manner. A number of issues are raised by this criterion. First, given the demonstrated strong effects of visual motion cues on a vehicle operator's behavior, adding physical motion cues may have little or no influence. Second, if physical motion cues are effective will full replication of physical motion be required. In other words, would the *scaling factor* of the simulator motion cues need to be at unity with the physical motion of the real vehicle under similar circumstances. In order to address these issues, we need to first understand how the human operator senses and processes the physical motion.

During vehicle movement, quite apart from the vection produced by optic flow, the vehicle operator is exposed to stimulation of both the vestibular system and the somatosensory system. The former evolved as means of providing balance and a gravito-inertial reference. It detects angular and linear *accelerations* of the head. The latter, the somatosensory system, is composed of the cutaneous mechanoreceptors that sense deformation (pressure) on the skin as well skin vibration and stretching. In addition to this tactile subsystem, receptors in the joints and muscles sense limb movement (kinesthesia) and limb position (proprioception). Both systems can provide a complex constellation of stimuli during vehicle operations. For that reason, their response characteristics are important if the goal is to design motion simulation technology with a high degree of perceptual fidelity.

THE VESTIBULAR SYSTEM

The primary nonvisual sense of self-motion is from the vestibular system, which is located as part of a bony labyrinth within the inner ear. The system evolved primarily for the detection of head movement, specifically head accelerations produced by changes in the head or whole body position. Postural stability, including a sense of a gravitational reference, is a central attribute of vestibular perception.

There are three principal axes of earth-referenced motion: (1) longitudinal (or X-axis), (2) vertical (or Z-axis), and (3) lateral (or Y-axis). Along these three axes of motion, there are two possible types of motion: (1) rotational or angular motion and (2) translational or linear motion (Figure 3.1). In general, vehicle motion can occur in any of these six degrees of freedom (DOF) and will often occur in more than one axis at a time. The human vestibular system is responsive to each of these six axes at any given time.

The vestibular system is composed of two subsystems (Figure 3.2). One subsystem consists of three semicircular canals arranged in rough approximation to the three earth-referenced axes of rotational motion (pitch, roll, and yaw). Rotational or angular movement around one of these axes by the head results in a movement of the endolymph fluid within a particular semicircular canal. Fluid movement in the canal, in turn, triggers hair cells in the cupola at the base of the canal. Nerve impulses

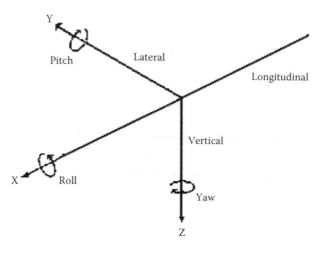

FIGURE 3.1   Axes of motion.

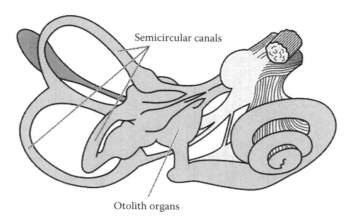

FIGURE 3.2   The vestibular system, including the semicircular canals and the otolith.

are then sent to the vestibular–ocular nuclei and then on to a variety of areas in the cerebellum (for motor control) and other areas.

For linear acceleration and gravity sensing, the vestibular systems also contain the otolith organ, which is composed of two membranous sacs: the saccule and the utricle. Hair cells embedded within calcium carbonate inside these sacs are triggered when linear accelerations are applied to the head. As the saccule and utricle are oriented at 90° to one another, they respond to vertical (saccule) and horizontal (utricle) linear accelerations independently. Although the semicircular canals rely on physical movement, the otolith functions as both a dynamic and a static system. The gravity sensing component of the otolith is the principal means by which we sense the earth vertical. The process is occurring even when the head is not moved. When aligned with the earth-vertical axis, gravitational forces apply a shearing action on the otolith and this, in turn, sends signals to the brain that the head is upright. As the head is moved away from the vertical axis, signals are sent to the brain to automatically correct for this deviation, that is, to regain postural stability. Lateral and longitudinal acceleration of the head, as well as vertical accelerations, affect the otolith and these forces play an important role not only in postural stability, but also in how a vehicle operator responds to forces resulting from operator control inputs or from disturbances external to the vehicle.

Both the semicircular canals and the otolith send their signals to vestibular and oculomotor nuclei and onward to a variety of sites in the brain. These include the thalamus for integration with other sensory information and onward from there to the cortex to support the perception of body orientation and movement. Additional neural connections to the brainstem reticular formation are sent to support changes in respiration and circulation and to the spinal cord for limb control that may be required of the new body orientation and position. One of the most important transmissions is the one that is sent to the cerebellum for motor control communication to control the eye and head position. The latter is essential to support the second function of the vestibular system and that is the stabilization of the visual image when the head is moved.

Image Stabilization

Whenever the head is accelerated, the visual image projected on the retinas of the eyes will also move. In order to allow for vision to work effectively during head movements, a method has evolved to stabilize the image. The vestibular ocular reflex (VOR) serves this function. During head movements, signals are sent from the vestibular system through the vestibular nuclei to the extra oculomotor nuclei and from there to the small ciliary muscles of the eyes. As it is driven by angular accelerations from the vestibular system, the VOR will function in complete darkness as well as when light is present. Contrast this with the optokinetic reflex (OKR), another reflex designed to stabilize visual images, but driven only by visual cues (retinal image blurring) when the visual image is in motion. In simulators with no physical motion present, no VOR is generated and only the OKR is available for image stabilization.

The sensitivity of the vestibular system is an important issue in the design of devices intended to simulate the motion of vehicles. As the vestibular system responds only to accelerations, motion devices must be assessed by their ability to provide *accelerations* at the appropriate time and duration relevant to the task and specific vehicle. It is not uncommon for the specification of a given simulator motion device to include some measures comparing the device's motion relative to that of the vehicle simulated. This assessment of rate or velocity of movement is used implicitly as means of assessing physical fidelity of a simulator's motion capability. However, measures of the amount of relative displacement of a motion system are not directly relevant in assessing a device's capability to produce perceptual fidelity because the vestibular system only responds to *accelerative* forces. The amount of maximum displacement of a motion system or its workspace, in turn, affects the amount of acceleration that can be provided per unit of time. The higher the level of acceleration required, the greater the workspace needed to accommodate that acceleration.

The sensitivity of a vehicle operator to vestibular stimulation can be measured in several different ways. The most common is the threshold measure. Thresholds to accelerations are measured by determining the point at which the human subject can detect whole body acceleration more than half of the time. Thresholds are usually expressed as averages for individuals and as a range of averages for groups of individuals. Thresholds can vary somewhat from one study to the next depending on the method and equipment used to measure them.

Angular Acceleration Thresholds

Thresholds to angular accelerations in complete darkness have been measured in yaw axis' rotational devices designed for that purpose. These thresholds have been found to range between $0.05°/s^2$ and $2.2°/s^2$ (Clark and Stewart, 1968a). There are other means for measuring thresholds in which a visual stimulus (e.g., a lighted visual cube) is fixed to the rotating device. The visual stimulus appears to move during angular accelerations in what has been termed the *oculogyral illusion* (OGI). The apparent visual movement of the OGI dramatically reduces thresholds to angular accelerations. With the OGI method, thresholds were reduced to a range of $0.05°/s^2$ to $0.18°/s^2$ (Clark and Stewart, 1968b). Using the same rotation device a year later, Doty (1969) found mean angular acceleration thresholds to the OGI ranging from $0.10°/s^2$ to $0.62°/s^2$ with thresholds lower with longer stimulus duration. A later study using the OGI method with a different rotation device found mean angular accelerations thresholds ranging from $0.02°/s^2$ to $0.95°/s^2$ (Miller and Graybiel, 1975). These authors note that three-fourths of those tested had thresholds below $0.20°/s^2$, whereas 90% had thresholds below $0.30°/s^2$ and 100% below $1.0°/s^2$. Clearly, when there is a visual stimulus present that covaries with the physical motion, there will be a dramatic reduction in threshold values. In addition, certain individuals who may have unusually high thresholds to angular accelerations will likely benefit more from such a visual stimulus than will others.

Linear Acceleration Thresholds

Thus far, studies of the responsiveness of humans to simple and single-axis angular accelerations have revealed both considerable stimulus sensitivity as well as an interindividual variability of some significance. Given the apparent mechanical simplicity of the vestibular system's semicircular canals, much more consistency would have been predicted. However, an analysis of the neuroanatomy of the vestibular system suggests that more processing of these signals may be involved than first thought. The otolith subsystem, which is principally responsible for sensations from linear accelerations, has been evaluated using a variety of devices for both horizontal and vertical movements and may also display variations in thresholds for the same reason as measures of angular acceleration.

One of the earliest attempts to define human response to linear motion was a review of 18 studies of thresholds to periodic linear motion by Gundry (1978). Periodic linear motion is exemplified by vehicle vertical movements in response to perturbations from disturbances such as air turbulence in the case of aircraft or surface roughness in the case of ground vehicles. A range of 0.014–0.25 m/s² was found in these studies. The range in measures reflects the variation in methodology as well as in the devices used.* A more recent study examining linear accelerations along all the three axes reveal somewhat lower thresholds overall but clear differences between the Z-axis (vertical) thresholds and the horizontal thresholds (Benson et al., 1986). In this study, mean threshold to vertical accelerations was 0.154 m/s², whereas the mean threshold to lateral (Y-axis) linear accelerations was 0.057 m/s² compared to 0.063 m/s² for longitudinal (X-axis) linear accelerations. The mean threshold to vertical acceleration, acceleration along the earth-vertical axis, was more than double that of the other two axes.

An interesting finding of threshold research is that the vestibular system does not appear to behave in the same way with regard to the presence of visual stimuli during linear motion events as it does with angular accelerations. Benson and Brown (1989) found similar significant reductions in thresholds to angular accelerations with OGI as found in the studies described earlier. However, thresholds to linear accelerations presented in darkness did not differ significantly from thresholds measured in the presence of a visual stimulus. The authors argue that this is evident that the two vestibular system subcomponents, the semicircular canals and otolith, do not interact with the visual system in the same way.

Task Characteristics and Physical Motion Perception

Estimates of the predictive validity of these sensory threshold data for the design of simulator motion are complicated further by the issue to task demand. As a vehicle operator has limited perceptual processing capacity, motion thresholds are likely to be affected by the demands of the task at hand. Clearly, the simple task of responding to movement in the absence of any other stimuli or even in the presence of a simple visual stimulus in a research device cannot be compared to the task demands of flying an aircraft or driving a car. It is perhaps not surprising therefore that research would reveal that the thresholds of motion perception would be affected by the task demand. Hosman and van der Vaart (1976) found a marked increase in motion thresholds of up to 80% for roll motion and up to 40% for motion in the pitch axis in the presence of an additional control or mental task. Gundry (1977) found that threshold to angular motion increased by 40% when a cognitive task (mental arithmetic) was added to a roll motion control task in a flight simulator.

---

* 9.81 m/s² = 1 g.

With regard to the use of threshold or other simple measures of operator response to physical motion, it is evident that signals form sensory receptors such as the vestibular system are affected by a variety of factors, including higher level cognitive processing restrictions imposed by varying levels of task demand. It is important for the designer to understand the complex nature of human motion perception that requires an understanding that goes beyond simple mechanical modeling of the vestibular system.

An additional concept for the study of motion perception was introduced by Gundry (1976) and has a particular relevance in any discussion of the higher-order perception of motion cues. When an operator of a vehicle initiates a particular maneuver the resultant physical motion forces are a product of the operator's behavior. This *maneuvering* motion, as it is called, confirms to the operator that an input has had an effect on the vehicle. Motion forces that occur as a result of *disturbances* to the vehicle, on the other hand, serve to alert the operator to a vehicle malfunction or an environmental event. In continuous manual vehicle control tasks, these disturbance motions are, by definition, outside of the operator's control. A physical motion cue that occurs as a consequence of vehicle control activity may be an essential part of that activity. Although disturbance motions need only to be suprathreshold and of sufficient duration to attract attention, maneuver motion as an integral part of control activity may require durations of motion of much greater magnitude. This means that the two types of motion would likely demand much different levels of physical displacement of the simulator and a much different design if only one of these is desired.

THE SOMATOSENSORY SYSTEM

The perception of the whole body motion is not limited to the vestibular system. As noted earlier, the neural pathways that extend from the vestibular system find their way to cutaneous and joint receptors involved in the process of maintaining balance. This somatosensory or *body sense* system is typically involved when the body is accelerated beyond certain limits and would be expected to be involved in many vehicle control tasks as a result of both maneuvering and disturbance motions. In this case, multiple receptor systems throughout the body may be involved.

One of the chief functions of the somatosensory system is in the support of postural control. For this reason, specialized neural pathways extend from both cutaneous mechanoreceptors as well as receptors in the muscles and joints. The latter are sensitive to muscle contraction and stretch (Dougherty, 2017). The speeds of neural transmission for pathways that enervate muscles are nearly twice that of cutaneous mechanoreceptors. This suggests a potentially greater influence of these muscles and joint receptors on proprioception (sense of limb position) and kinesthesia (sense of movement) and possibly a much greater impact on postural control. The latter neural transmissions also proceed along a separate pathway from cutaneous mechanoreceptor stimulation allowing localization of sensory information and ultimate integration in the cerebral cortex.

Cutaneous Mechanoreceptors

The effects of vehicle motions are often transmitted to the skin of the operator's buttocks, thighs, and lower back as well as other skin receptors such as the plantar sensors on the base of the feet. The functions of these mechanoreceptors and their response characteristics have been reviewed by Johnson (2001). Four receptor types have been identified within the dermis. First, the Merkel disk receptor has a slow-adapting* response to skin deformation. The nerve signal activity increases as a linear function of the extent of that deformation. The second receptor type, the Meissner corpuscle, is a rapidly adapting receptor that is insensitive to skin deformation but very sensitive to skin movement. The third receptor, the Pacinian corpuscle, is a rapidly adapting receptor specialized for low amplitude (e.g., 10 nm at 200 Hz) vibration. The fourth receptor type, the Ruffian corpuscle, is a slow adapting receptor that is sensitive to the stretching of skin and to the direction of the forces, which produce the skin stretch. The Pacinian corpuscles are located deeper in the skin than all other mechanoreceptors.

Although the operating characteristics of these receptors are known, much less is known about the distribution of the cells around the body. High densities of these receptors are known to exist in the hands and fingers in which receptor density ranges from 100 to 150 per cm^2. However, receptor density decreases dramatically in the areas of the lower back and legs with densities less than a tenth of that found in the hands and fingers. This is reflected in the low point localization discrimination ability in areas of low receptor density (Weinstein, 1968). A study by Kennedy and Inglis (2002) found a total of only 104 receptors of all types in the sole of the foot and of these, most were of the fast adapting type. The authors suggest that the dominance of this type of receptor point to the plantar region of the foot as essential in maintaining balance.

The low density of mechanoreceptors suggests that large areas of the lower body extremities involved in motion response need to be stimulated if the simulator motion cueing is to be successful. As we will see in the remainder of this chapter, this is the design strategy followed by early attempts to simulate somatosensory perception in vehicle simulators.

The density of cutaneous mechanoreceptors in areas of the body thought to provide somatic cues is one issue that needs to be addressed in simulating somatosensory stimulation. But other issues also need to be considered. One issue is the thickness of epidermal tissue, which overlays and protects the underlying dermis in which the receptors reside. As noted earlier, the plantar region of the foot not only has an overabundance of fast-adapting receptors, but the sole of the foot has skin with a much thicker epidermis than other parts of the body. Likewise, the lower back has an epidermal layer closer in thickness to that of the face and hands. Studies of the thickness of the epidermis such as Whitton and Everall (1973) should be consulted by device designers for more details.

---

* Slow adapting receptors are best suited to transmit impulses from long-lasting stimuli, fast adapting respond only to the onset and offset of stimuli.

In addition to receptor density and skin thickness, proximity to muscle and bone also affects how skin receptors are likely to function. As vehicle operators are typically seated upright, linear accelerations in the vertical axis (g_z forces) will be imparted primarily to the buttocks. Most notable is the effect of the ischial tuberosities (also called the *sitting bones*) on the distribution of pressure across the buttocks. Pressure is maximized at points 0.04–0.06 m from the body's center line (sagittal plane) then rapidly drops off after this area (Hertzberg, 1955). Much of the sensation of g_z forces as well as vibrations imparted to the vehicle operator's body through the seat pan will thus be concentrated in this area. In addition, lateral accelerations, such as those in ground vehicle operations around curves, will result in weight shifting, which will be sensed in this area of the buttocks as well as the outer area of the thighs.

The cutaneous mechanoreceptors in the skin aid in detecting limb motions in combination with the vestibular system as part of a system to maintain the posture as well as transmitting vibrations and g_z forces to the vehicle operator. In addition, more recent studies implicate these receptors in the detection and position of limb movement and in the forces applied to the limbs and other extremities (Dougherty, 2017). There are other receptors within the muscle fiber, such as the muscle spindle, which detect contraction and extension of skeletal muscles and thus the movement of the associated body part such as the limbs and neck. In addition, the Golgi tendon organ that attaches to the connection of the muscle fiber and the tendon detects forces placed on the muscle and transmits the resulting nerve activity to higher centers of the brain. These neural transmissions communicate the position of extremities, including the head, neck, and limbs as well as their movement to the higher brain levels, including the cerebellum for motor activity and the somatosensory cortex. Attempts to simulate specific forces on the head and neck will involve these specialized receptors.

Vibrations

Unlike the motions discussed so far, vibrations imparted to the vehicle operator's body are characterized by rapid oscillations, usually sinusoidal, about a specific neutral point. Again, most of these are occurring as longitudinal or lateral sinusoidal, periodic, or aperiodic whole body vibrations. These vibrations can occur in vehicles at frequencies below 1 Hz and as high as 30 Hz or more at amplitudes well below the threshold for linear accelerations. Vibrations occur in vehicles for a variety of reasons, both normal and abnormal. Normal vibrations occur from the engine, wheels, suspension, body frame, and other vehicle equipment. These vibrations are *expected* by the experienced vehicle operator as a result of many hours of contact with controls and other components of the vehicle such as the seat and cockpit floor. The *absence* of normal vehicle vibrations may also be an important indicator of potential vehicle malfunctions in real-world vehicle operations. Likewise, abnormal vibrations, those that occur at unusually higher or lower frequency or amplitude than the operator has experienced, can signal problems.

Equipment failures, including vehicle structure or engine components, can cause severe vibrations in both aircraft and ground vehicles. Vibrations can be also experienced by the vehicle operator when encountering unusual environmental events.

These include low-level air turbulence in aircraft as well as vibrations due to surface irregularities in taxiways and runways. In ground vehicles, irregularities in road pavement and in terrain surfaces are the most commonly occurring, environmentally induced vibrations.

Vibration Thresholds

Vehicle operators are most likely to sense vibrations through contact with control surfaces due to the increased density of cutaneous mechanoreceptors in the skin of the hands and fingers. Pacinian corpuscles in the dermis of the skin in this area are also closer to the surface of the skin than other areas, such as the feet, due to thicker epidermal layers in those areas. Transmission rates of vibration will likely be higher as well because other areas, such as the seat and seat back materials will attenuate vibrations to some extent. Absolute threshold to vibration stimuli can vary depending on how and where on the body the assessments are made. A study evaluating the absolute threshold of vibration when the subject grasped a cylinder 32 mm in diameter approximates the grasping behavior of vehicle operators. In this study, vibration thresholds decreased linearly from amplitudes of 14 μm at 10 Hz to 5.6 μm reaching a minimum of 0.03 μm at 150 Hz and 200 Hz then increasing (Brisben et al., 1999).

Whole body vibration (WVB) threshold is, however, a potentially more useful measure as the vehicle operators entire body is likely to be affected by vehicle vibrations and not just the fingers and hands. (However, note the potential for a low cost means of vibration cueing using solely the operator's controls as input sources). Parsons and Griffin (1988) measured WVB thresholds for seated subjects for vertical vibrations between 2 and 100 Hz to be 0.01 m/s². Sensitivity to vibration, not surprisingly, is greatest at the hands for frequencies above 100 Hz (Morioka and Griffin, 2008).

PHYSICAL MOTION SIMULATION TECHNOLOGY

In any of the discussion on vehicle simulation, optimizing physical fidelity means that the simulator must move just as the vehicle moves. Optimizing motion cueing or perceptual fidelity means that the simulator needs to provide the operator with the same perceptual experience as would be experienced in the real vehicle. Attempting to duplicate the physical motion or even the motion cueing of vehicles, particularly aircraft vehicles, has proven to be very difficult. This is primarily, but not exclusively, due to the need to provide the sustained linear accelerations experienced in both ground and air vehicles. The first serious attempt to duplicate this type of motion began with research simulators used in the field of aerospace medical research and testing. Aircraft, especially high-performance aircraft, can induce high levels of linear accelerations: positive and negative. In modern fighter aircraft, sustained g_z up to and beyond 9 times the level of gravity can be achieved.

To achieve such levels of sustained linear accelerations would require motion simulation devices measured more in square kilometers than in square meters if the motion were to be produced in a straight line. For example, linear accelerations of only a modest 1 m/s² for 5 s would require 12.5 m of displacement and for 10 m/s² (around 1 g), a displacement of at least 128 m is required. The answer for sustained high-level, linear accelerations was the centrifuge. Simulating these linear accelerations by means of a centrifuge simplified the design problem. A centrifuge consists of an arm attached to and rotating about, a central point. A cockpit is attached at the end of the arm. In order to provide g_z, the cockpit is rotated to a position in which the longitudinal axis of the pilot's body is aligned with the arm. If the head of the pilot is positioned toward the center of the centrifuge, positive g_z will be produced. Alternatively, if the pilot's head is positioned toward the outside of the centrifuge away from the center point, negative g_z will be produced. Lateral accelerations can be produced by a perpendicular alignment of the pilot's longitudinal body axis to the centrifuge arm. To date, only the centrifuge is capable of simulating high levels of sustained linear accelerations on the ground. The principal use of the centrifuge has been for medical and perceptual research in high-performance aircraft and for space flight. Although valuable for research, particularly for high sustained g_z, the centrifuge is cost-prohibitive for routine training and testing purposes.

Alternative Designs

A number of other attempts to simulate linear and angular accelerations comparable to those experienced by vehicle operators have been developed over the past few decades in response to a number of research needs in both aircraft and ground vehicle design. (A comparison of the wide range of working space of some of these simulators is shown in Table 3.1). For research on vertical linear accelerations, for example, motion simulators such as NASA's Vertical Motion Simulator (VMS) has been developed to research motion effects on pilot-controlled behavior primarily in the earth vertical axis. The VMS has a vertical envelope of approximately 24 m (78.7 ft), which allows accelerations of up to 22 m/s^2 (about 0.75 g_z).[*] The VMS also provides about 12 m (39 ft) of additional translational motion allowing accelerations of ±4.9 m/s^2 in the longitudinal axis and ±4 m/s^2 in the lateral axis. Roll, pitch, and yaw angular accelerations can be provided up to ±229°/s^2 (Aponso et al., 2009).

TABLE 3.1
Displacement Capabilities of a Variety of Motion Platforms

	Linear Displacement (m)			Angular Displacement (°)		
	Vertical	Lateral	Longitudinal	Pitch	Roll	Yaw
NASA–VMS	±9.2	±6.2	±1.2	±18	±18	±2.4
NADS	±0.62	±9.8	±9.8	±25	±25	±330
SIMONA	±1.15	±1.15	±1.15	+24.3/−23.7	±25.9	±41.6
DESDEMONA	±1.0	±4.0	±4.0	360	360	360
NASA B747	±0.68	±0.68	±0.68	±32	±32	±32
Renault CARDS	±0.30	±0.30	±0.30	±30	±30	±30

[*] The height was based on studies that determined that at least ±20 ft of displacement would be needed to meet motion cue fidelity requirements during the approach and landing phase of flight (Bray, 1973).

Other research simulators have employed large-scale motion platforms in order to address specific vehicle research needs. For ground vehicle simulators, the National Advanced Driving Simulator (NADS) was recently developed to research issues in the passenger car and truck operations. As with the NASA VMS vertical motion envelope, NADS extends the lateral and longitudinal motion envelope to allow for linear accelerations to be more similar to those of ground-based vehicles in these axes. NADS can provide linear accelerations of ±0.6 g in the lateral and longitudinal axes and ±1 g_z in the vertical. Angular accelerations in the pitch, roll, and yaw access are provided at a maximum of $\pm120°/s^2$ (Chen et al., 2001). Maximum WBV frequency is 20 Hz with a displacement of ±0.5 cm.

Recent innovations in motion platform technology have evolved to provide more effective cueing in multiple motion axes. The Desdemona simulator (TNO, The Netherlands) is a hybrid configuration incorporating a gimbaled cab allowing sustained angular accelerations up to a maximum of $90°/s^2$. The cab is mounted on a rotating track arm, which provides centrifugal forces up to 3 g. The cab can also move laterally along the track arm for up to 8 m (26 ft) (maximum 5 g) to provide additional accelerations. Finally, the cab is suspended vertically for 2 m (6.6 ft) to allow for accelerations up to 0.5 g. The advantage of this type of configuration allows a complex combination of linear and angular accelerations that are not available with other motion systems (Feenstra et al., 2007; Bles and Groen, 2009).

One of the more innovative approaches, and potentially more cost effective, is the use of modified industrial robot arms. One of these, the MPI Cybermotion Simulator at the Max Planck Institute, mounts a cab at the end of a 6 DOF robot arms with acceleration limits ranging from $33°/s^2$ to $128°/s^2$. The robot arm itself is mounted on a track with up to 9.9 m (32.5 ft) of travel (Nieuwenhuizen and Bülthoff, 2013).

These motion platform systems described earlier have been designed for a variety of research purposes, including, but not limited to, the study of motion cueing on operator behavior. These systems are relatively rare due to their high costs. Much more common in both research and in training and testing is the Stewart platform. The system was designed with both economy of space and lower cost in mind and has become the standard in motion platform system. The NASA 747 simulator, a Federal Aviation Administration (FAA) certified full mission simulator is typical of the capabilities of these Stewart platforms (Sullivan and Soukup, 1996). The motion platform displacement capability of the simulator compared to other research simulators is notable.

The Stewart Platform

Without question, the most common motion platform system in use today is some variation of the Stewart platform[*] (Stewart, 1965). The Stewart platform is a multiaxis, synergistic motion platform typically consisting of six drive legs (hexapod) supporting a platform on which a cab is mounted (Figure 3.3). The drive legs typically have maximum extensions of 0.5–3 m with the midpoint of the extension used as a neutral point. The typical configuration has extensions of ±0.6 m for training platforms and ±1.2 m or more for simulators used in research. The design is an

---

[*] Sometimes called the hexapod.

FIGURE 3.3   The Stewart platform.

elegant solution to the problem of providing motion in all the 6 DOF within the smallest possible platform workspace. The very limited motion envelope that the design offers eliminates the need for large facilities, thereby dramatically reducing the facility-related costs as well as the cost of the system itself. The lower cost of the Stewart platform has made it the system of choice in deployment of large numbers of simulators used in training and testing. The Stewart platform has gained rapid popularity primarily because it fulfilled the low cost requirement that vehicle operators desired. Moreover, it filled the criteria, particularly of pilots, that favor at least some motion cueing, however minimal, to none at all. This *user acceptance* element of simulator motion systems does not mean that the system is effective in providing motion cueing, only that it may support a certain level of subjective realism. Indeed, its effectiveness as a motion device in the training regime is still very controversial.

Motion Cueing Algorithms

An element of any platform motion design that is critical to effectiveness is the motion cueing software that controls how the motion platform performs. In a simulator motion platform, the platform is initiated either through the operators control input or from the external disturbances commands. Before these commands are sent to the platform, however, they are input through a vehicle dynamics module, which computes the effect the command will have on the vehicle's physical motion. However, because the motion has a restricted workspace, the commands from this module must be filtered in order to assure that the commands to the platform actuators will not exceed their physical limits. Two factors at this stage become important for the perceptual fidelity of the system. The first factor is the motion gain or a scaling factor that is applied to the command. This factor adjusts the command signal by some fractional amount, so that the resultant simulator motion force will be of some proportion to the real-world vehicle motion. This proportion is usually less than 1.0 (or unity gain) and more than 0.2. The second factor is the phase distortion, also called the break frequency, which automatically eliminates commands that are below a predetermined frequency. This blocks low amplitude signals that would require much more displacement than the platform workspace would allow. A phase distortion of the command signal will be much higher for platforms with very restricted workspace, such as the Stewart platform, and much lower for large-scale platforms such as the NASA–VMS. Even with identical motion gains, higher phase distortion is likely to result in poorer perceived fidelity in motion platforms. A careful balance between the motion gain and the phase distortion is required to maximize the performance of a motion platform.

One additional factor in the design of motion platforms is noteworthy. In all motion platforms, the requirement exists for the system to return to its drive actuators to a neutral position. Typically, this is at the midpoint of the actuators full-drive extension. In order to return the actuator while the simulator is in operation, the motion drive software must assure that the return of the simulator platform occurs without providing a (false) motion cue to the operator. To avoid this, the drive software must command the return of the platform with subthreshold levels of motion. That is, a level of motion that is undetectable by any operator of the device. This is where the careful application of motion perception threshold data becomes most critical.

MOTION PLATFORM CUEING SYSTEM EFFECTIVENESS

Although the design of simulator motion systems is impressive, it is the performance of the simulator *user* that is of paramount importance in establishing perceptual fidelity. Of the behaviors of most interest in that regard, it is the ability of the operator to control the simulated vehicle in a manner comparable to, if not indistinguishable from, that which the operator exhibits in the real vehicle under comparable circumstances. The problem with these criteria is that we rarely have performance data from the field on a sufficiently large, representative sample in order to determine with some confidence the baseline performance that could serve as a target for simulator comparison.

Methodology

For this reason, platform motion system effectiveness is usually evaluated in one of two ways. The first way is to evaluate the motion system in the simulator itself. This is done by comparing the performance of two or more groups of experienced operators each performing the identical task in an identical simulated operational environment: one group receiving physical motion cueing and one group receiving no motion cueing. If operator behavior performance is better with motion cueing than without, it is evident that some level of improved perceptual fidelity has been provided. Different variations of this experimental design have been employed.

For example, some axes of motion may be disabled or the motion gain or scaling factor for a given motion axis may be altered so that the motion cues may be more or less similar to that of the real vehicle. This may be a simple matter of turning off all motion cueing (a fixed-base condition) or limiting or constraining some aspect of the motion system. Finally, a comparable group of operators is exposed to a full-motion cueing condition, or more accurately, the maximum amount of motion a simulator motion system can provide. In any event, all groups involved must be comparable, so that the effects of these motion cueing conditions are not confounded by some aspect of the operator such as experience.

This *in-simulator* method has the advantage of experimental control over the conditions of testing, especially the control of variations in simulator motion cueing that are of interest. Task conditions can be held identical between or among groups and the assignment of individuals to groups themselves more tightly controlled. However, this method is not a direct measure of how experience in the simulator will affect operator behavior in the actual vehicle. It is only a measure of the extent to which, if any, the simulator motion cueing is effective in changing the operator's behavior. In addition, it is important to understand that the differences that may be shown by this method can only reveal motion cueing effectiveness within the limits of motion provided by the simulator platform or related motion cueing device. This may or may not be a full measure of the extent of motion cueing effectiveness in the real vehicle, but rather a measure of whether some degree of perceptual fidelity is available from the system. In the absence of any motion cueing effects in such a study it can only be concluded that, given the limits of motion cueing available to the operator by the particular system, there is no evidence of any effects on behavior. The in-simulator method is especially useful for assessing the impact of physical motion cueing on the behavior that is already fully formed in the operator. This method of the in-simulator assessment method, sometimes called *reverse transfer*, provides more direct evidence of motion cueing effectiveness as it avoids the confounding influence of the skill acquisition process.

The second method of evaluating motion cueing is the training transfer study. The transfer of training study is the type most often cited when reviews of motion systems are discussed. This type of simulator evaluation consists of training a vehicle operator in the simulator and then evaluating that individual's performance later in either the actual vehicle that was simulated or an acceptable surrogate (e.g., the same or another simulator). The transfer of training study has historically been used as a means to operationally evaluate simulators to determine their suitability and effectiveness. Measuring the operator's performance in the real vehicle following simulator training is considered the *gold standard* of simulator effectiveness by training organizations.

Selected Studies Criteria

It is important to note that only those studies that met certain criteria have been selected here for discussion. First, the fidelity of the perceptual experience can only be determined reliably by evaluating how the system affects the behavior of *experienced* operators as these operators have developed a fully formed task-relevant, perceptual cue utilization capability. It is these individuals that possess the internal representations, which relate to the perception of a motion cue to a specific control behavior. Second, only those studies that used an adequately sized sample of operators were selected in order to achieve sufficient test power to allow for accurate statistical inference to take place. Finally, only studies that provided objective evidence for improvements in perceptual fidelity. This requires an objective measure of behavioral changes, such as changes in pilot control behavior. Subjective measures of realism, handling quality ratings, comparative workload ratings, and similar measures, although useful in themselves, are not considered definitive. This is due to the influence that individual biases have on subjective ratings and assessments that may reduce the validity of these data with respect to simulator design components, especially controversial areas such as motion cueing.

There is perhaps no issue in the field of simulator design that has engendered more controversy than the effectiveness of motion cueing systems. In vehicle simulators, motion cueing systems, particularly large motion platform systems, add a dramatic degree of face validity to a simulator by their size and complexity. It is understandable then that an operator using such devices would be inclined to believe that they are necessary and effective deign component of the simulator system. This reason alone justifies the emphasis on objective behavioral data as definitive in the measurement of motion cueing effectiveness.

The effectiveness of these platforms and related motion cueing systems has been the subject of research for at least 40 years. Over this time period, dramatic technological changes have occurred particularly in computer technology and these changes determine whether results from one system are replicated in another system years later. The studies also have used a large variety of methods, motion cueing systems, vehicle types, and simulator design features. Some restrictions on the inclusion of studies here have been employed in order to help clarify a complex issue.

First, only those simulator studies that provide an external visual scene to the operator are included here. A simulator with no visual scene capability may force the operator to use motion cues to an extent that is not representative of how these cues would be used in the real vehicle. Indeed, those who have conducted such studies have found this to be the case (Jacobs, 1976). In some early studies, the display FOV is much less and the display resolution is poorer than that available to an operator in the real vehicle or even in current simulators. These issues should be considered when weighing the validity of the data.

Second, there are substantial differences in how different vehicle categories respond to operator control inputs and disturbances from external factors. Airborne and ground vehicles, for example, differ in the number and interrelationships that exist among axes of motion as well as in how they respond to forces in the real world. In addition, within vehicle types there are significant differences in vehicle mass and in inherent stability. In some cases, subdivisions of these two vehicle categories are necessary for the same reason. The inclusion of different vehicle types here is intended to reveal the complexity of the issue of motion platform effectiveness within the vehicle simulator market.

Finally, the *type* of motion platform system will play a role in how effective it will be in providing motion cues to the operator. As noted earlier, the designs of these systems can have a marked effect on how much linear or angular acceleration can be imposed on the operator in relation to that which would be available in the actual vehicle. Principal among these motion platform characteristics is displacement, that is, the distance the platform can physically move in any axis of motion. To help understand this more clearly, Figure 3.4 shows the relationship between acceleration and displacement over time for selected linear vehicle accelerations.

The sample of acceleration levels are representative of motions typically encountered by large transport aircraft of the type used in commercial airline or military air cargo operations and in ground vehicles such as passenger cars. In this example, the vehicle is assumed to be accelerated from a position of rest for a period up to a maximum of 9 s. If the position of rest is the neutral position of the simulator's drive leg and the simulator has ±0.6 m of displacement, then the platform can be expected to deliver at most 2 s of linear acceleration at 0.3 m/s^2 (assuming no portion of the drive leg extension is limited for safety purposes). Accelerations at higher levels, such as 1 m/s^2 (about 0.1 g), would be sustained for less than 1 s. As research studies have only demonstrated human responsiveness at durations of 0.5 s or higher, it is likely that a duration of 0.5 s of exposure is the minimum requirement in order for motion accelerations to be effective. However, it is also possible that longer durations of acceleration are needed in order for a given acceleration to exceed the threshold when the operator is experiencing high task demand. In this case, the actual effectiveness of a motion platform might be significantly impaired even though the platform's performance specifications would suggest otherwise.

The relationship between time and acceleration means that motion platforms, even advanced research platforms, have only a very limited space in which to provide

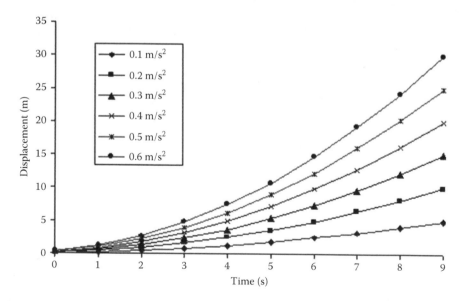

FIGURE 3.4 Minimum platform motion displacement required for linear accelerations in large transport aircraft and ground vehicle simulation. Accelerations are from starting velocity of zero.

motion cues for simulated vehicles. For linear accelerations, even for relatively small levels of accelerations above the perceptual threshold, the amount of platform displacement required becomes very large, very quickly.

Fixed Wing Aircraft

The category of fixed wing aircraft includes all aircraft, military and civilian, in which the wing of the aircraft is fixed to the fuselage. This is in contrast to rotary wing aircraft such as helicopters where the wing, that is, the source of aerodynamic lift, is free to rotate. These two distinct categories are needed to distinguish between aircraft with an inherent aerodynamic stability, fixed wing aircraft, and those which are inherently unstable, helicopters. In addition, propulsion systems on fixed wing aircraft may be attached to the wings at points far from the fuselage. At these points, engine failures can have dramatic maneuvering and disturbance motion effects on motion sensations experienced by the pilot. It is for this reason that evaluations of motion cueing effectiveness usually employ the engine failure scenario.

Early studies of the effectiveness of motion platform cueing were largely aimed at military aircraft of one sort or another. This is understandable given the heavy dependence on flight simulators for training military pilots. In a simple compensatory tracking task in a fighter aircraft simulator equipped with a Stewart platform, the pilot tracking error performance was reduced significantly with motion platform cueing (Ashworth et al., 1984).

A study of the effects of motion platform cueing in a large military transport aircraft research simulator found positive effects on an experienced pilot performance (Showalter and Parris, 1980). Pilot performance in three tasks: (1) engine failure during takeoff, (2) precision turns, and (3) landing with wind shear. Each was evaluated in the presence or absence of platform motion cueing. Note that the first and third tasks involve disturbance motions, whereas the second does not. Performance was enhanced by the presence of motion platform cueing in the first and third task, but not in performance in precision turns.

The interpretation of the results of both these studies needs to be tempered by the fact that both used visual display systems of limited FOV (<50° h) and with low resolution (9 arc min). Nonetheless, these studies provided initial support for the use of motion platforms in fixed-wing aircraft flight simulators in simple tracking tasks or in which disturbance motion was present.

Additional studies have been conducted using more modern devices and, most important, with the wide FOV visual systems with higher resolution (3 arc min/pixel or less). A study by Lee and Bussolari (1989) evaluated the effectiveness of motion cueing using a modern Stewart platform system in a simulator (B727-200) certified by the FAA as a full-flight mission simulator. Experienced commercial airline pilots flew tasks involving both maneuvering and disturbance motions. Motion platform cueing conditions varied from full 6 DOF, reduced to 2 DOF, and then to a very limited motion cueing condition simulating a variety of special effects such as runway rumble and flap buffet. In the latter condition, the platform displacement was limited to only 0.63 cm (0.25 in.) of travel. The pilots were unaware of what type of motion was employed in each condition. The study revealed no effects of motion cueing on any pilot behavior in any of the maneuvers. In all conditions, pilots believed the motion to be comparable to that of the actual aircraft despite the large differences in motion cueing capability that was presumably available. The importance of this finding is that, not only did large variations in motion platform capability have no discernible effect on control behavior but that the pilot's subjective assessment of motion realism could be influenced by very little motion. Notably, the motion special effects were consciously felt by the pilots, so physical validity was immediate and obvious.

The absence of any influence of platform motion on pilots in this study is particularly important as the flight simulator used was certified by the FAA for the training and evaluation of airline pilots. The study would undermine the argument that motion platforms are required for at least this type of aircraft. When compared to the results of Showalter and Parris (1980), the differences are dramatic. Two possible alternative explanations are available. First, the aircraft vehicle dynamics of the two aircraft were quite different resulting in different aircraft responses to disturbance motion. Environmental and equipment disturbance motions elicited very dramatic behavior from the aircraft in the Showalter and Perris study when compared to the much more docile behavior of the aircraft in the Lee and Bussolari (1989) study. Second, there were differences in the availability of external visual cues between the studies. The second study provided much more of a visual FOV and visual detail of the external world and therefore allowed potentially much less reliance on the use of physical cues to control the aircraft. Third, and perhaps most important, the research simulator of Showalter and Perris provided much more lateral motion cueing from the extensive (30.5 m) lateral track platform.

An alternative view is that the modern Stewart platform used in these training simulators significantly understates the motion cues that would normally be available in the real aircraft. The Stewart platform has a very limited workspace, which necessitates substantially reduced motion displacements if all motion axes are to function fully. Motion algorithms deliberately reduce the amount of motion gain that the platform actuators will receive, so that only a fraction of the actual computed aircraft motion will actually be generated. This is done to assure that the commanded accelerations do not push the drive legs beyond the platform's available space. The particular implementation of motion washout in training devices does not take into consideration any differences in the effects that the motion may have for different axes of motion under different tasks.

Motion Platform Design and Aircraft Type

It is a fact that different aircraft types impose different motion forces for similar events. A more recent study examined motion cueing of a Stewart platform when one axis of motion was allowed a higher performance than would be normally available (Burki-Cohen et al., 2003). A flight simulation of a large, multiengine aircraft (B747-400) was employed. The simulator had a modern wide display FOV and was certified by the FAA for pilot training and evaluation. The basic motion platform characteristics were the same as the study by Lee and Bussolari (1989) but the platform motion algorithm was altered. Lateral motion cueing (linear accelerations) were amplified by altering the existing motion gain and phase distortion of the platform, so that the lateral motion cues would be about 0.85 of the actual aircraft lateral motion from the original scale factor of about 0.2 that was used in the simulator's certified configuration. The result of these alterations was to substantially sacrifice the pitch and the yaw motion cueing while significantly increasing the motion gain at a level much higher than this type of motion platform would normally provide.

The task for the pilots was to detect and respond to an engine failure on takeoff. The lateral motions produced by the enhanced motion platform proved to be effective in aiding the detection of the engine failure as measured by the more rapid and effective rudder pedal response to the disturbance motion. Clearly for this particular task, the lateral accelerations resulting from the engine failure provided an important alerting function. The results of this study corroborate the early study by Showalter and Perris (1980) at least with respect to this particular task. Also note that the cueing effect was produced within the very limited displacement available to a Stewart platform.

Essentially, the same task was evaluated in a simulator of a twin turboprop commuter transport by Go et al. (2000). The twin turboprop has only a fraction of the gross takeoff weight of the typical jet airliner. A FAA-certified simulator with a Stewart platform was used, but the platform motion system was unaltered. Motion cueing was effective in improving the pilot's yaw motion control of the simulated aircraft when engine failure was introduced during takeoff.

Differences in the type of aircraft simulated are likely to impact the effectiveness of the Stewart motion platform design. Principally, the effect of wing-mounted engines on the above-mentioned aircraft types compared to the fuselage-mounted engines in Lee and Bussolari (1989) study was the main factor. The lateral and yaw forces are significantly greater than those of aircraft such as the B727 when engine failures occur on takeoff. An additional factor was the significant difference in the mass of the two aircraft, which contributed to much higher accelerations in lighter twin turboprop aircraft in a very similar task. Finally, the lateral translational motions of large transport aircraft during this type of maneuver require motion platform designs which may differ significantly from those that support the yaw rotational motion of small aircraft.

A similar method of enhancing the accelerations available to different axes of Stewart platform motion has been done by others (Beukers et al., 2009). In this study, roll, pitch, yaw, and the lateral axes of motion were tested individually and in combination with a business jet flight simulator. A single task, the decrab maneuver, was chosen for evaluation. The decrab maneuver is executed just prior to landing when the pilot is attempting to land by orienting the aircraft's heading (*crabbing*) into a crosswind. In order to land without damaging the landing gear, the pilot must decrab the aircraft before landing. This is a task that will produce maneuvering rather disturbance motions. The results of the study showed a reliable effect of the combined motion cueing in yaw and lateral axes for rudder pedal activity with a motion scaling factor of only 0.6.

More recently, a study evaluating the characteristics of the Stewart platform for civil aviation transport use was conducted using the NASA–VMS (Zaal et al., 2015). This study altered the characteristics of the NASA–VMS to emulate the Stewart platform when limited to 0.3, 0.6, and 0.85 scale factor of the actual motion of the simulated aircraft, a large twin jet transport. These adjustments to the Stewart platform were determined to be of medium physical fidelity by the investigators. Experienced pilots flew three tasks under each of these platform variations and one control motion condition of unrestricted NASA–VMS motion (0.8 scale factor, but very low phase error). The latter was classified as high fidelity. Only a difference in one performance measure (maximum pitch rate) in one of the three tasks (stall recovery) was found among the three Stewart platform conditions. Significant differences were found between the Stewart platform and the NASA–VMS in all three tasks.

The limitations of the Stewart platform as it is typically configured for large, civil airline transport aircraft are clearly demonstrated in the above-mentioned studies. For fixed wing aircraft, without the modifications suggested by the studies discussed earlier, the Stewart platform is unlikely to provide the perceptual fidelity needed for motion cueing to be effective in the flight critical engine failure on takeoff task. Some benefits to pilot performance did appear when light turboprop and business jet aircraft were simulated during landing maneuvers. The benefit of this motion system appears for large aircraft to be in the simulation of special effects motions, such as runway rumble, that elicit favorable ratings of realism by users but have themselves no effect on control behavior.

More important, the experimental modifications to the Stewart platform demonstrated that motion cueing could be effective in this type of aircraft, not just for disturbance, but for maneuvering motion cues as well at least for some aircraft. Although the effect sizes were small in these studies, they were nonetheless statistically reliable. The importance of motion cueing in disturbance-related tasks is supported. The evidence for maneuvering motion cueing is still very limited. Most importantly, the studies showed that simulator motion cueing with unity gain, a one-to-one correspondence to real aircraft motion (full physical fidelity) is not necessary to achieve perceptual fidelity.

Rotary Wing Aircraft

Fundamental to the issue of motion cueing has been its role in vehicles which have poor or marginal stability. Modern fixed-wing aircraft have inherent stability, that is, they tend to maintain a given state unless otherwise disturbed and tend to return to that state when the disturbance ends. Rotary wing aircraft, that is, helicopters, are notable for their lack of inherent stability. Even with modern stability augmentation systems, training, testing, and research will continue to be conducted in helicopters without these augmentation systems in use. Unlike fixed-wing aircraft, helicopters can perform tasks such as hovering, vertical takeoff, and other tasks that produce unusual motion sensations for pilots. These tasks place a high demand on pilot control skill.

Although disturbance motion is notable for its value as an alerting cue, the use of motion cues in helicopters is less obvious. The central argument put forth by Young (1967) is that motion cues in vehicles with marginal or poor stability (e.g., helicopters) serve to aid control by providing nonvisual information that *leads* visual motion when the vehicle motion reaches the outer limits of control. The nonvisual cues thus might be thought of as a limiting or a compensatory factor in closed loop vehicle control for aircraft with stability issues. The absence of this compensatory motion cueing in a simulator would mean that pilots would have a much greater difficulty controlling a helicopter in a fixed-based simulator than in one which provides this compensatory cueing through a motion cueing device.

A series of studies conducted by Schroeder (1999) using the NASA–VMS simulator exemplify the complexity of motion cueing for helicopter simulators. The simulated helicopter was configured as an attack helicopter (AH-64) flown by experienced pilots performing a variety of motion sensitive tasks, including hovering and precision yaw control. Yaw, roll, pitch, and lateral motion cueing was evaluated. One of the main findings was the apparent irrelevance of yaw rotational motion to the performance of these tasks. Instead, the most important motion cueing came from the lateral translational motion. The author argued that the visual cues provided by the external visual scene display were more than adequate in providing yaw motion cues so platform cues were, in effect, made irrelevant. Motion platform scaling of 0.8 for translational motions and 0.4 for rotational motions are supported in agreement with modeled data from Sinacori (1977).

In an additional study of an altitude control task, Schroeder (1999) determined that vertical translational motion cues were also found to be very important. However, reductions of the scale factor from 0.8 to 0.6 were recommended for vertical translational motions when compared to other translational motions. That the scale factor for platform motion could be reduced substantially suggests that other sensory information, such as somatosensory stimuli resulting from gravitational-induced tactile pressure to the pilots' buttocks, plantar region of the feet, back of the thighs, and other areas of the body were involved.

A later study using the SIMONA research simulator, a Stewart platform, attempted to address the lack of yaw rotational motion cueing in Schroeder (1999) study (Ellerbroek et al., 2008). The study found an interaction between the yaw rotational and the lateral translational motion cueing. The latter interaction revealed that yaw motion cueing improved performance but only in the presence of lateral motion. Analyses of the differences between the two studies revealed that the only plausible reason for the different results was due to the differences in the sample of pilots used. Although both the studies used the same number of experienced helicopter pilots, the Schroeder study used exclusively helicopter *test* pilots. It is possible that unlike the pilots in the Ellerbroek et al. study, the helicopter test pilots were able to extract more information from the visual scene provided to them even though this scene was of somewhat lower resolution in the Schroeder study. This would have allowed them to generate more visual lead compensation, an important element in the control of a low stability vehicle such as a helicopter.

Ground Vehicles

Simulation of ground vehicles includes not only passenger cars, trucks, and buses, but also military ground vehicles such as tanks. Railed vehicle simulators also exist, such as trains and light rail but are not normally provided with motion cueing devices. In ground vehicles, the issue of motion cueing has not been as controversial as it has in the design of aircraft simulators. However, unlike aircraft simulators, ground vehicle simulator costs are much more constrained. It is far easier to justify the high cost of aircraft simulators, including the additional cost of motion cueing because aircraft simulators are often replacement devices for aircraft often costing millions or tens of millions of dollars. It is common to find ratios of 10:1 or higher between the cost of an aircraft simulator and the actual aircraft it simulates. This high cost ratio can be justified as a means to reduce the high cost of aircraft operations as well as eliminating the potential damage or loss of an aircraft due to accidents.

The issue of whether motion platform devices are necessary for ground vehicles to achieve high perceptual fidelity, however, has engendered the development of large-scale motion platforms for use in research simulators in government, academic, and commercial organizations. The motion platforms in these devices, as with aircraft simulator motion platforms, vary dramatically depending on the focus of the research for which they have been designed. However, the limited motion accelerations in the vertical axis and the more extensive motion accelerations possible in both longitudinal and lateral axes of ground vehicles are commonly reflected in their design. Many commercial vehicle manufacturers also employ large-scale research simulators for a variety of reasons.

One of the reasons for the development of large motion platforms is the investigation of lateral and longitudinal linear motion effects on driver performance under a variety of task and vehicle conditions. As the duration and intensity of linear accelerations are dependent upon motion platform displacement, the cab of the simulated vehicle is sometimes mounted on tracks which can extend many meters in either direction. Angular accelerations are provided by variations of the Stewart platform and typically provide only limited motion cueing in the pitch, yaw, and the roll axes.

Linear accelerations along the longitudinal axis are normally controlled by the driver with braking and throttle inputs. A study of motion cue effectiveness in this axis was conducted using the Renault dynamic simulator equipped with a Stewart platform (Siegler et al., 2001). Motion gains on some axes were altered in order to enhance longitudinal platform motion. Experienced drivers were required to decelerate in order to stop as close as possible to a sign. Repeated trials were conducted in order to examine the effect learning might have on the use of motion by drivers. The overall effect of motion cueing was to prevent the drivers from excessive and unrealistic levels of deceleration. Most importantly, drivers in the fixed-base condition increased their braking force for those over trials, whereas the braking force receiving motion cueing remained constant. The drivers in the fixed-base condition were apparently repeatedly adjusting their control strategy to compensate for the lack of nonvisual motion cueing.

The Siegler et al. (2001) study also examined the effects of combined lateral and roll motion cueing were compared to the combined longitudinal and pitch on performance in a cornering task. Both these conditions were then combined and compared to no motion. Drivers in the task tended to track further away from the side of the road when lateral motion cueing was present but not in other motion conditions. Although the effect was small, a mean track deviation of only 20 cm, it was statistically reliable. When longitudinal motion cueing was present, linear velocities of the vehicle were also lower suggesting that the drivers are deliberately altering their speed control strategy when motion cueing is present.

The use of lateral acceleration motion cues by drivers was also found in a study by Savona et al. (2014) using the SHERPA simulator. This simulator consists of a Stewart platform and an X–Y platform and was used to evaluate driver response to lateral translational cues (produced by a combination of translation motion and platform tilt) and the roll and yaw cues in a slalom course. The severity of the slalom course was systematically varied, so that the lateral accelerations produced by driving would be 1, 2, or 4 m/s². Motion scale factor was then systematically varied for each of these acceleration levels. The speed of the car was fixed at 70 km/h. For the performance measure, steering wheel reversal rate, the results revealed that the best scale factors were 1.0 for lateral motion and 0 for both the roll and yaw in the easiest slalom course (1 m/s²); 0.5 for lateral motion and 1.0 for both roll and yaw for the more difficult slalom course (2 m/s²); and 0.25 for lateral motion and 1.0 for both the roll and yaw acceleration in the most demanding course (4 m/s²). In this study, the drivers could not reduce vehicle speed to reduce lateral accelerations. Analyses of steering suggested that motion gains below 0.2 should not be used for lateral motion. Driving accuracy for the more difficult slalom courses (higher acceleration levels) was better when the motion gain was reduced. Roll and yaw motion gains of 1.0 produced the best driving performance, but only at higher levels of acceleration.

Motion platform cueing is also effective when the driver must cope with crosswind disturbances. In a study of crosswind effects on driver heading control using the VIRRTEX simulator, equipped with a Stewart platform, and directional control of the vehicle was improved when the driver was provided with lateral motion cues (Greenberg et al., 2003). Most important, due to the ability of this simulator to provide motion-scale factors up to 0.7 of the vehicle's actual lateral accelerations, the authors could systematically manipulate the scale factor in the study. The authors found that a scale factor of 0.5 of the actual vehicle's linear acceleration optimized the drivers' performance. No further gains in performance were found beyond that scale factor. Greenberg et al. (2003) also found motion cueing to be effective in normal highway driving when the driver was distracted by other tasks unrelated to driving. When operating a handheld phone during driving, the driver loses the external visual cues normally associated with the task. When platform motion cueing was provided, lane violations attributable to this task were reduced nearly eight-fold when compared to the fixed-base condition. Clearly, nonvisual motion cues can be of value even under otherwise normal conditions when visual motion cues from the roadway are absent.

However, motion cueing does not appear to be effective in altering the performance of the experienced drivers in more routine driver tasks. For example, speed and distance control behavior of drivers are unaffected by the presence of motion cueing in a vehicle following task (Columbet et al., 2008). This was true despite the fact that the drivers preferred the presence of motion cueing and thought it to be more realistic.

Somatosensory Motion Cueing Effectiveness

The motion platform systems discussed so far have been designed with the belief that providing vestibular motion cues is necessary to achieve at least some level of motion fidelity. These systems have the capability to provide motion sensations well above the thresholds for either linear or angular accelerations, though not for sustained periods of time. Studies of platform motion effectiveness suggest that this technology may meet at least some of the objectives of perceived fidelity, though the effectiveness varies with the vehicle type and the task required for the vehicle operator. Motion platforms are costly and this high cost has led to the development of alternative means of motion cueing.

Motion platforms provide not only the vestibular sensations but also somato-sensory motion cues. They do so principally by stimulating the cutaneous mechanoreceptors in the skin and the receptors in the muscle and joints as the platform moves. This element of motion cueing could conceivably be simulated by alternative, nonplatform devices, which could provide the same force cueing as motion platforms, but at a much lower cost. Similarly, such devices might also provide the vibration cues that have been found to be so useful in creating a sense of realism in simulator users.

The early history of somatosensory motion cueing was primarily targeted at the development of devices that could provide the sensations of high g_z forces experienced by fighter pilots. These forces have broad effects on the pilot's body but they are specific forces that force the pilot's head, limb, and torso into the aircraft seat during combat maneuvering.

One of the first attempts at simulating these forces with somatosensory cueing devices was the helmet loader. The device attached cables to the pilot's helmet that pulled down on the pilot's helmet as high g_z were encountered in the simulator. As the pilot reduced the g_z loads, cable pressure on the helmet was reduced (Ashworth and McKissick, 1979).

Of all the somatosensory cueing devices that have been devised, the g-seat is probably the most studied and the most likely to find its way into the modern simulator in one form or another. The g-seat was originally designed as a possible alternative or as an additional motion cueing device to that of the Stewart platform. The initial designs were intended for use in fighter aircraft for simulating somatosensory cues associated with high-g maneuvers; hence, the term *g* seat. The g-seat consists of a number of pneumatically or hydraulically driven seat panels in the seat pan, thigh, and seat back areas. When the angular or linear accelerations are encountered the appropriate seat panels are activated providing pressure cues. The resultant pressure sensation associated with the simulated maneuver is intended to provide sensations similar to those experienced in the real vehicle. A seat belt tensioning device is typically used to provide additional restraint and tactile cueing when the seat is activated under positive linear accelerations. In aircraft seats, the seat belt is actually a complex 4-point seat harness, which restrains the pilot securely during flight. For ground vehicles, a more conventional 3-point system might be used. A variety of g-seat designs have evolved over time, descriptions of which can be found in Kron (1975) and, more recently, Sutton et al. (2010) and deGroot et al. (2011). The latter two articles describe *dynamic* seats in which the entire seat may move a very limited distance.

The design of the g-seat, as essentially a tactile pressure system, relies on a belief that some simulated vehicle motions will exceed the pressure threshold of the cutaneous mechanoreceptors in each of the active areas of the device. Note that, for a variety of reasons, the thresholds will not be the same in all areas of the body. For example, we know that the area of the ischial tuberosities are likely to result in lower thresholds than other areas as there is less body mass between these boney protuberances and the seat itself. Similarly, the thigh area of the legs has a larger amount of soft flesh, which may increase the sensation threshold to pressure. This means that the somatosensory cueing thresholds for vertical accelerations are likely to be lower than the lateral accelerations simulated by the g-seat. Finally, differential pressures can be provided to the seat pan and back to simulate roll, yaw, and pitch cues. For example, a roll to the right could be simulated by increasing the pressure in the left seat pan area with concomitant reduction in pressure to right seat pan area.

The g-seat can take advantage of the fact that the cutaneous mechanoreceptor neurotransmissions vary linearly with increases in skin indentation. Thus, increasing pressure on the skin surface through increased displacement should result in an increasing perception of linear or angular motion. As mechanoreceptors are most sensitive to low frequency (<50 Hz) vibration as well as to pressure in the upper (epidermal) layer of the skin, g-seat components do not need to have extensive panel displacement in order for the device to be effective.

Early studies of the g-seat in a fighter aircraft found it to be effective in simple, compensatory tracking tasks. In these tasks, a reticle (gun sight) is overlaid on a target aircraft and the pilot attempts to maintain the reticle on the target as it maneuvers. Target tracking was found to be improved by the presence of g-seat motion cueing (Ashworth et al., 1984). In a study of pilot performance in executing precision turns in large transport aircraft, the g-seat improved its performance in roll and pitch control but not in engine failure recognition and recovery or in a landing task (Showalter and Parris, 1980). The use of the g-seat to augment cueing of a fixed-base simulation transport landing task was not found to be effective (Parrish and Steinmetz, 1983).

G-seat evaluations in helicopter operations have found performance improvements in vertical maneuvers doing hover (White, 1989) but not during shipboard landing tasks requiring precise lateral maneuvering (Westra et al., 1987; Chung et al., 2000).

These early studies reveal the utility of the g-seat in simple tracking tasks and in more complex tasks in which the emphasis is on cueing in vertical translation motion. It is much less effective in tasks in which lateral or longitudinal motion cueing is needed.

Dynamic seats, also called motion seats, have been less successful in affecting operator performance when employed in driving simulators. Only very limited effects of a dynamic seat were found on braking behavior, for example. A reduction in initial deceleration levels when compared to a fixed base condition was found in a driving simulator study (deGroot et al., 2011).

The g-seat may be useful in providing limited vibration simulation to the vehicle. As noted earlier, the study by Lee and Bussolari (1989) found that the realism provided by only very limited movement of the motion platform was equivalent to that provided by the full 6 DOF of platform motion. As these special effects and random disturbance motions resulting from environmental factors are important for subjective realism, perhaps the g-seat may provide a low-cost alternative to motion platforms in this case. Even though g-seat panel displacements are typically very limited (<2.5 cm), their available displacement should be sufficient to exceed the median vibration threshold of 0.01 m/s^2 (Parsons and Griffin, 1988). It is not expected that the g-seat would be capable of high-amplitude vibrations such as those found in heavy ground vehicles (Kaman, 2004) or in significant aircraft turbulence that can easily exceed 1 m/s^2.

SUMMARY OF PHYSICAL MOTION EFFECTS

It is a common belief that a vehicle simulator must physically move, just as the real-world vehicle moves, in order for the simulator to be considered high in fidelity. Specifically, that the device must replicate the accelerative motion, both linear and angular, and must do so in all axes of motion. A device that provides less than full motion or only partial motion is not considered to be high in physical fidelity and thus should be considered less valid in training, evaluation, and research.

A contrasting view is that fidelity should be measured objectively by its effect on user behavior and not on measures of physical fidelity. Second, that physical motion is only needed to support specific control activity under specified conditions. Third, certain kinds of motion cueing (e.g., disturbance cueing) need not match real-world motion in order to provide information value. Motion cueing can, in fact, be effective at only fractional levels of real-world intensity.

Review of the relevant literature that has evaluated vehicle simulator motion cueing systems with experienced users does not support the need for full physical fidelity, either in the requisite unity scaling factor or in the six axes of motion. The value of physical motion in control activity depends on a variety of factors. These include the type of vehicle, the vehicle dynamics, and the task being performed. Variations in task demand appear to affect the perceptual threshold of vehicle operators such that previously established thresholds in low demand environments may misrepresent threshold levels in the higher demand environments of vehicle operation. This, in turn, suggests that limited workspace motion devices, such as the Stewart platform, are likely to be much less effective than originally thought. Indeed, in some cases, the Stewart platform may be providing only limited subjective realism, but no actual benefit to operator control activity at all (Lee and Bussolari, 1989).

Advanced research simulators with large-scale motion platform devices have shown that motion cueing can contribute to operator control activity for some vehicle types and for some tasks. The required scaling factor or gain for this motion cueing to be effective, however, depends on the vehicle being simulated. For fixed-wing and rotary-wing aircraft simulators, this scaling factor appears to be quite high, ranging from 0.6 to 0.85. For ground vehicles the scaling factor range is much lower: 0.4–0.7. These scaling factor values become very important in establishing the most cost-effective workspace for motion platform devices.

An important finding with regard to *how* the operator is actually using motion cues is important in determining whether a motion device is likely to be of value for a particular vehicle or task. It is evident that one of the most important aspects of physical motion is its role in disturbance events such as those caused by equipment failures or environmental phenomena. Disturbance motion cues play a significant role in alerting the operator to a potentially dangerous event. Due to the faster neural transmission time of physical motion and the absence of a need for an orienting response (i.e., *looking*), it can reduce the reaction time of events when compared to visual cues alone.

Motion cues also appear to play a role in what might be called a limiting cue in that the operator can use the physical motion as a means to reduce certain excessive vehicle behavior in simple tracking tasks. It can also serve as a means to control excessive speed in curves by providing lateral accelerative (g_y) forces when safe speeds are being exceeded. Finally, physical motion appears to function similarly in the longitudinal axis to avoid excessive decelerations in braking behavior.

In marginally stable vehicles, such as helicopters, physical motion serves to provide the pilot with anticipatory cues, which aid in reducing the instability. This lead compensation may have applications in other control areas such as turbulence or in off-road applications for vehicles. In these cases, the operator may use these motion cues to compensate for disturbances to the vehicle.

It should be noted that only studies that have compared the behavior of experienced operators with and without physical motion present were considered in order to assess whether, and under what conditions, physical motion might contribute to perceived fidelity. There is sufficient evidence to support the claim that physical motion does indeed contribute to perceived fidelity in both aircraft and ground vehicle simulators. However, this does not mean that the effect is universal for all vehicle types and classes nor does it apply to all vehicle tasks. Moreover, the physical motion produced by limited workspace devices such as the Stewart platform may not be adequate in either duration or in intensity. In all cases save one, motion effects were found in only advanced research simulators with extensive motion capabilities. In the one exception (Burki-Cohen et al., 2003), extensive modifications were required to the Stewart platform to enhance lateral motion capability. This modification effectively nullified two of the remaining motion axes. Finally, the issue of whether physical motion is necessary for operator training has yet to be addressed. The studies reviewed addressed only experienced operators and not those who have yet to achieve proficiency. However, it should be immediately apparent that a simulator that lacks physical motion, which would otherwise support optimum or even near optimum operator performance, is likely to impact the degree to which optimum operator performance can be achieved in a simulator.

SUMMARY

Providing physical motion in vehicle simulators is one of the more difficult engineering tasks and arguably the most controversial of all vehicle simulator components. This chapter reviewed the fundamentals of human motion sensing systems as well as the basic limitations and capacities of vestibular and somatosensory motion perception. The innovative technologies applied to the problem of motion simulation were reviewed, including those used by the most advanced research simulators. Although these simulators have shown some effects of motion simulation on operator performance, motion platforms of the type used in training, and testing of vehicle operators outside of research facilities have only been able to show the effects if the platform design was dramatically altered. The implications for providing perceptual cues associated with physical motion in vehicle simulators are described with particular reference to the importance of perceptual fidelity.

REFERENCES

Aponso, B.L., Beard, S.D., and Schroeder, J.A. 2009. The NASA Ames vertical motion simulator—A facility engineered for realism. *Royal Aeronautical Society 2009 Flight Simulation Conference*, June 3–4. London, UK.

Ashworth, B.R. and McKissick, B.T. 1979. Effects of helmet-loader cues on simulator performance. *Journal of Aircraft*, 16, 787–791.

Ashworth, B.R., McKissick, B.T., and Parrish, R.V. 1984. Effects of motion base and G-seat cueing on simulator pilot performance. NASA Technical Paper 2247.

Benson, A.J., Spencer, M.B., and Stott, J.R. 1986. Thresholds for the detection of the direction of whole body linear movement in the horizontal plane. *Aviation, Space and Environmental Medicine*, 57, 1088–1096.

Benson, A.J. and Brown, S.F. 1989. Visual display lowers detection threshold of angular, but not linear, whole body motion stimuli. *Aviation, Space, and Environmental Medicine*, 60, 629–633.

Beukers, J.T., Strroosma, O., Pool, V.M., Mulder, M., and van Paasen, M.M. 2009. Investigation into pilot perception and control during decab maneuvers in simulated flight. *AIAA Modeling and Simulation Technologies Conference*, August 10–13. Chicago, IL.

Bles, W. and Groen, E. 2009. The Desdemona motion facility: Applications for space research. *Microgravity Science and Technology*, 21, 281–286.

Bray, R.S. 1973. A study of vertical motion requirements for landing simulation. *Human Factors*, 15, 561–568.

Burki-Cohen, J., Go, T.H., Chung, W.W., Schroeder, J., Jacob, S., and Longridge, T. 2003. The effects of enhanced hexapod motion on airline pilot recurrent training and evaluation. *Proceedings of the 12th International Symposium on Aviation Psychology* (AIAA-2003-5678). April, 2003.

Brisben, A.J., Hsiao, S.S., and Johnson, K.O. 1999. Detection of vibration transmitted through an object grasped in the hand. *Journal of Neurophysiology*, 81, 1548–1558.

Chen, I.D., Paplis, Y., Watson, G., and Solis, D. 2001. NADS at the University of Iowa: A tool for driving safety research. *Proceedings of the 1st Human-Centered Transportation Simulation Conference*, November 4–7. Iowa City, IA: The University of Iowa.

Chung, W.W.Y., Perry, C.H., and Benford, N.J. 2001. Investigation of effectiveness of the Dynamic seat in a Blackhawk flight simulation. *AIAA 2001 Simulator Technology Conference and Exhibit*, August 6–9. Montreal, Canada.

Clark, B. and Stewart, J. 1968a. Comparison of methods to determine thresholds for perception of angular acceleration. *The American Journal of Psychology*, 81, 207–216.

Clark, B. and Stewart, J. 1968b. Comparison of sensitivity for the perception of bodily rotation and the oculogyral illusion. *Perception and Psychophysics*, 3, 253–256.

Columbet, F., Dagden, M., Reymond, G., Pere, C., Merienne, F., and Remeny, A. 2008. Motion cueing: What is the impact on the driver's behavior? *Driving Simulator Conference*, Monaco, January 31–February 1.

deGroot, S., deWinter, J.C.F., Mulder, M., and Wieringa, P.A. 2011. Nonvestibular motion cueing in a fixed-base driving simulator: Effects on driver braking and cornering performance. *Presence*, 20, 117–142.

Doty, R.L. 1969. Effect of duration of stimulus presentation on the angular acceleration threshold. *The Journal of Experimental Psychology*, 80, 317–321.

Dougherty, P. 2017. *The Somatosensory System*. Houston, TX: University of Texas. www.neuroscience.uth.tmc.edu (accessed May 14, 2017).

Ellerbroek, J., Stroosma, O., Mulder, M., and van Passen, M.M. 2008. Role identification of yaw and sway motion in helicopter yaw control tasks. *Journal of Aircraft*, 45, 1275–1289.

Feenstra, P.J., Wentink, M., Rozo, Z.C., and Bles, W. 2007. Desdemona: An alternative moving base design for driving simulation. *Proceedings of the Driving Simulator Conference, DSC 2007, North America*, September. Iowa City, IA.

Go, T.H., Burki-Cohen, J., and Soja, N.N. 2000. The effect of simulator motion on pilot training and evaluation. *AIAA Modeling and Simulation Conference*, August 14–17. Denver, CO.

Greenberg, J., Artz, B., and Cathey, L. 2003. The effect of lateral motion cues during simulated driving. In *DSC North America Proceedings*, October 8–10, Dearborn, MI.

Gundry, A.J. 1976. Man and motion cues. *Paper presented at the Third Flight Symposium on Theory and Practice in Flight Simulation*. London, UK.

Gundry, A.J. 1977. Thresholds to roll motion in a flight simulator. *Journal of Aircraft*, 14, 624–631.

Gundry, A.J. 1978. Thresholds of perception for periodic linear motion. *Aviation, Space, and Environmental Medicine*, 49, 679–686.

Hertzberg, H.T.E. 1955. Some contributions of applied anthropology to human engineering. *Annals of the New York Academy of Sciences*, 63, 616–629.

Hosman, R.J.A.W. and van der Vaart, J.C. 1976. Thresholds of motion perception measured in a flight simulator. *12th Annual Conference on Manual Control, NASA TM X-73*, May. p. 170, Washington, D.C.: National Aeronautics and Space Administration.

Jacobs, R.S. 1976. Simulator cockpit motion and the transfer of initial flight training. Doctoral Dissertation, University of Illinois at Urbana-Champaign.

Johson, K.O. 2001. The roles and functions of cutaneous mechanoreceptors. *Current Opinion in Neurobiology*, 11, 455–461.

Kaman, S. 2004. Vibration in operating heavy haul trucks in overburden mining. *Applied Ergonomics*, 35, 509–520.

Kennedy, P.M. and Inglis, J.T. 2002. Distribution and behavior of glabrous cutaneous receptors in the human foot sole. *Journal of Physiology*, 538, 995–1002.

Kron, G.J. 1975. Advanced simulator for undergraduate training: G-seat development. *Air Force Human Resources Laboratory* (AFHRL-TR-75-59).

Lee, A.T. and Bussolari, S. 1989. Flight simulator motion and air transport training. *Aviation, Space, and Environmental Medicine*, 60, 136–140.

Miller, E.F. and Graybiel, A. 1975. Thresholds for the perception of angular acceleration as indicated by the oculogyral illusion. *Perception and Psychophysics*, 17, 329–332.

Morioka, M. and Griffin, M.J. 2008. Absolute thresholds for the perception of fore-and-aft, lateral, and vertical vibration at the hand, the seat, and the foot. *Journal of Sound and Vibration*, 314, 357–370.

Nieuwenhuizen, F.M. and Bülthoff, H.H. 2013. The MPI cybermotion simulator: A novel research platform to investigate human control behavior. *Journal of Computing Science and Engineering*, 7, 122–131.

Parrish, R. and Steinmetz, G. 1983. *Evaluation of g-seat augmentation of foxed vs. moving base simulator for transport landings under two visually imposed runway width conditions*. NASA TP 2135. April.

Parsons, K.C. and Griffin, M.J. 1988. Whole body vibration perception thresholds. *Journal of Sound and Vibration*, 121, 237–258.

Savona, F., Stratulat, A.M., Diaz, E., Honnet, V., Houze, G., Vars, P., Masfrand, S., Roussarie, V., and Bourdin, C. 2014. The influence of lateral, roll and yaw motion gains on driving performance on an advanced dynamic simulator. *SIMUL 2014: The Sixth International Conference on Advances in System Simulation*.

Schroeder, J.A. 1999. Helicopter flight simulation platform motion requirements. *NASA TP-1999-208766*, July.

Showalter, T.W. and Parris, B.L. 1980. The effects of motion and g-seat cues on pilot simulator performance of three piloting tasks. NASA Technical Paper 1601, January.

Siegler, I., Reymond, G., Kemeny, A., and Berthoz, A. 2001. Sensorimotor integration in a driving simulator: Contributions of motion cueing in elementary driving tasks. In *Proceedings of Driving Simulation Conference*, September 21–22.

Sinacori, J.B. 1977. The determination of some requirements for a helicopter flight research simulation facility. NASA-CR-152066. Moffett Field, CA: NASA–Ames Research Center.

Stewart, D. 1965. A platform with six degrees of freedom. *The Proceedings of the Institution of Mechanical Engineering*, 180, 371–386.

Sullivan, B.J. and Soukup, P.A. 1996. The NASA 747-400 flight simulator: A national resource for aviation safety. AIAA, pp. 374–384.

Sutton, P., Skelton, M., and Holts, L.S. 2010. Application and implementation of dynamic motion seats. *Interservice/Industry Training Simulation and Education Conference.*

Weinstein, S. 1968. Intensive and extensive aspects of tactile sensitivity as a function of body part, sex and laterality. *The Proceedings of the First International Symposium on the Skin Senses,* Tallahasee, FL: Florida State University.

White, A. 1989. G-seat heave motion cueing for improved handling in helicopter simulators. *AIAA Flight Simulation Technologies Conference and Exhibit*, Boston, MA.

Whitton, J.T. and Everall, J.D. 1973. Thickness of the epidermis. *British Journal of Dermatology*, 89, 467–476.

Westra, D.P., Shepard, D.J., Sherri, A., and Hettinger, L.J. 1987. Simulator design features for helicopter shipboard landings II. Performance experiments. U.S. Naval Training Systems Center Technical Reports, July. Technical Report 87-04153.

Young, L.R. 1967. Some effects of motion cues on manual tracking. *Journal of Spacecraft and Rockets*, 4, 1300–1303.

Zaal, P.M.T., Schroeder, J.A., and Chung, W.W. 2015. Transfer of training on the vertical motion simulator. *Journal of Aircraft*, 52, 1971–1984.

4 Manual Control—Force Perception

INTRODUCTION

All vehicles require some type of control input and, as such, control system behavior is a critical element in the design of a vehicle simulator. The control system itself can vary widely in complexity and in cost. Regardless of its design, all control systems serve as a means by which the vehicle operator communicates with the vehicle and, through force feedback along with other sensory input, the means by which the vehicle communicates with the operator. For initial training of manual control skills in which the operator needs to acquire effective control strategies, incorporation of force feedback may be necessary. Manual control in many vehicles such as aircraft, rail vehicles, and ships is becoming less frequent as more automation is used to supplant the human operator. This does not mean, however, that manual control fidelity in simulation design is less important. In fact, because the operators are using manual skills much less often in these vehicles, manual skill retention will decline over time. The degradation of these skills necessitates retraining and reevaluation in the real vehicle or in a vehicle simulator. Indeed, the risk of skill deterioration due to automation remains a concern throughout the aviation, particularly civil aviation, as many complex vehicle control operations, which previously required manual control are now nearly fully automated.

The consequences for those operating highly automated vehicles are that skill maintenance will now be largely in simulators, not in the real vehicle. Although this has been true in the past for many skills that are too dangerous to practice in the real vehicle (e.g., aircraft engine failure on takeoff), increasingly it is the simulators that will become the primary means of maintenance for many routine manual skills because it is often too costly to use the real vehicle for training. The emphasis on the vehicle simulator as a primary venue for manual skill maintenance raises the issue of the perceptual fidelity of the simulator control system to a higher level.

As with the problem of visual imagery and physical motion, perceptual fidelity in manual control is dependent on how the experienced operator perceives the control input forces and the control force feedback required to control the vehicle. In addition, the displacement of the control device itself contributes to the perception of the control process. This applies to both continuous control input devices, such as the automobile steering wheel, and to discrete control input devices such as switches, knobs, and other devices, including touch input controls.

Most of the critical control inputs in manual control are carried out without actually looking at the control. Rather, the primary proximal feedback is through physical contact with the control. That contact is typically through manual manipulation by the hand in the control wheel and control stick inputs and by the feet in pedal controls. Additional forces can be applied by the operator through the forearm and shoulder muscles in the case of steering wheel and stick type controls and the use of leg and foot muscles in the case of pedal controls. Generally, most control skills do not require large force requirements due to the advances in hydraulically assisted, electronically assisted, and other power-assisted technology introduced into modern vehicles. These technologies are common in large aircraft and in large ground vehicles, which previously required substantial physical strength for manual control but now require relatively light control forces. In some cases, as in fly-by-wire or drive-by-wire control systems, the operator's control input is simply a signal sent from a transducer to a computer. In this case, the control forces and any control displacement are artificially created. This addition of artificial force feel and feedback and displacement to a control device in which none previously existed reflects the importance of perceptual cues for manual control.

Although this chapter focuses on the perceptual cues of force, force feedback, and control displacement, it should be remembered that manual control usually involves a complex interaction between the manual control input and the visual and the non-visual processing of the consequences of that control input. Thus, the manual control forces and feedback are taking place in the context, and often simultaneously with, other perceptual processing.

THE FORCE PERCEPTION SYSTEM

Manual control in conjunction with sensory systems such as vision is the means by which humans interact with their environment. This perceptual-motor control system allows humans to physically respond to stimuli and to manipulate and apply forces to objects in their environment. Perceptual-motor control is essential to human adaptation and therefore for survival.

Postural stability, locomotion, and other perceptual-motor systems have an inherent, preexisting neural architecture that allows for rapid development and refinement in infancy and is well developed by childhood. It is important to understand the nature of this architecture as it influences how the perceptual-motor system functions and what components are involved. It is particularly important to note that the systems, once initiated, tend to operate automatically, free of conscious interference.

Newer perceptual-motor skills learned during a lifetime leverage the basic neural architecture of survival skills such as postural stability. Details of the basic architecture are described by Willingham (1998) and will be summarized here. The description of this architecture is intended to review the aspects of this architecture that are of importance in simulator design and evaluation. This is especially true of those aspects of perceptual-motor control that are susceptible to conscious manipulation. In simple perceptual-motor control tasks in which only feedback from the physical displacement of the control device is available, the task goal is consciously selected by the operator. This task goal is a specific displacement parameter stored in the dorsal prefrontal cortex. This area of the brain is the locus of high-level decision-making and problem-solving in humans. In this instance, the task goal and its plan for completion are processes subject to conscious alteration prior to completion. The task goal and its completion criteria are quite simple, for example, move the control lever to the left about 2 cm. At this level, such complex task goals, subgoals, and associated operational parameters are possible and allow conscious alteration as well as verbal description. Also note that the higher cortical areas are responsible for task management, specifically task prioritization, in order to allow operators to manage task load more effectively.

Once the task goal and parameters are set, signals are transmitted from the dorsal prefrontal cortex to the motor cortex. The primary function of the motor cortex is to initiate the actual motor control act and to monitor the feedback from the control action. This is done by sending signals to lower level motor control areas such as the cerebellum. At this level, motor control acts are no longer subject to conscious manipulation or alteration.

The feedback from this motor act is necessary in order to determine whether the action plan task goal has been met. In many control acts, there are usually several potential feedback sources. There may be a visual verification that the control act is completed by simply viewing the control device displacement or, as is more common in vehicle control, the physical force feedback sensed by the tactile receptors combined with the proprioception of displacement. These multiple perceptions are sent to an area of perceptual-motor integration in the posterior parietal cortex. Here, all the available perceptual inputs are processed. This may include the application of the weightings regarding the reliability or relevance of each input to the completion of the task at hand. The results of this processing are transmitted back to the motor cortex, which sends commands to the cerebellum and then to the spinal cord. Finally, signals are transmitted to the appropriate musculature.

In some motor control acts, more than one act is performed in sequence. For example, automobile braking typically involves releasing the pressure on the accelerator pedal before applying pressure to the brake pedal. In experienced drivers, this becomes a highly automated sequence of control actions solely with reference to proprioceptive and force cues. Although susceptible to conscious monitoring and alteration, the sequence of perceptual-motor control acts appears to be localized in an area of the brain termed the supplementary motor area (SMA). Signals are then transmitted on to the basal ganglia, to the thalamus (for sensory integration), and then looping back to the SMA. Closed-loop control acts, such as steering wheel and control yoke inputs, are executed at this low level of brain activity.

The important element in this neural architecture is the separation of high-level goal and motor control sequencing representations from their low-level executions. In the earlier framework, only the high-level goals are consciously available. For designers, reliance on subjective descriptions regarding the forces or displacement required for a particle control device to execute a particular task is likely to be unreliable. Thus, the forces and displacement applied to a steering wheel in an automobile driving task, such as maintaining lane position, are not accessible to conscious description. Moreover, objective assessment of driver performance needs to be conducted within the context of the entire task. Thus, evaluating the control forces and displacement required of the steering wheel in control lane position needs to be conducted with the presence of real-time, visual, auditory, and other cues that are normal components of the lane positioning task. Attempting to evaluate control force and displacement cues in isolation is likely to yield inaccurate measurements as it would not include the prioritization and other sensory integration that normally occurs in the task. Thus, the design of control device forces and displacement should take place within the entire process of perceptual and perceptual-motor processing.

The perceptual-motor control that a vehicle operator exercises becomes highly integrated and highly automated with experience. The operator has conscious control over the selection of the particular task goal such as lane position, speed, and so on. The second portion of the process, perceptual-motor integration, extracts the sensory feedback from the environment necessary for the performance of the task. These include visual, auditory, physical motion, proprioceptive, and tactile forces that are associated with motor control movement. The repetition of control access during the training and initial operational experience of the vehicle operator results in an increasingly fine-tuned motor control act. At this point, conscious access to the different inputs in the control act is difficult, if not impossible to do without a substantial impact on performance. Conscious attention to a low-level input such as control position and its effects is indeed a characteristic of the novice operator and is reflected in high subjective mental workloads.

The perceptual-motor integration process is particularly important for the simulator design process because simulator designs often involve either the elimination or the degradation of one or more perceptual elements that normally support real vehicle operation. The visual scene may lack adequate detail to support the distance perception or inadequate resolution to allow detection of targets in the visual scene. For the experienced operator, the result may be an inability to perform tasks or it may result in an attempt to select different inputs in order to complete the task. Both of these invariably have a negative effect on operator performance and may or may not affect subjective realism.

Force perception, specifically the proprioceptive-tactile perception that results from the operation of a control device, follows the perceptual-motor integration process as the operator moves the control to achieve the desired result. The force perception involved in the control device operation is an integral element of the perceptual-motor integration process. The forces applied to the control are driven, and continuously updated by, the feedback from an array of other perceptual inputs. Adjusting the forces required to move a control device and to achieve a particular task goal involves a continuous, closed-loop process of adjustment and readjustment until the task goal is met. For a driver's lane position task on a straight road, the adjustments are relatively few to achieve the desired result when compared to those that are required when the road is curved.

The highly learned, even over learned, motor control processes involved in the use of vehicle control devices means that experienced vehicle operators will necessarily anticipate that the same or similar forces will be applied to controls in the vehicle simulator. The question as to how well vehicle operators can discriminate different levels of force required to operator control device becomes relevant here. This is especially true for critical control devices such as aircraft control yolks and automobile steering wheel systems. Essential control devices such as these require development of high levels of perceptual fidelity in simulators and deserve more attention than they have historically received.

Static versus Dynamic Forces

In order to achieve high levels of perceptual fidelity in simulated vehicle control, the perception of forces required to move a control device should be well within the operator's ability to discriminate differences in the control forces required for the vehicle simulator as compared to the real vehicle. That is, the control force differences, if any, between the real vehicle and the simulator should not be detectable by the vehicle operator.

In measuring the force that is necessary to move control devices the designer needs to discriminate between passive and active forces. All control devices have an inherent mass that is quite apart from the components of the vehicle they are connected to or the environment in which the vehicle operates. At the level of human motor control, the mass of the control device is used by the vehicle operator to determine the force required to displace it. The force required to overcome the mass of the control device is static. That is, the force required is the same regardless of task conditions.

In the real world, of course, the control devices are connected to other vehicle components and, by extension, to the environment in which the vehicle operates. Thus, not only the mass of the control device itself but that of the other components within and outside the vehicle must be overcome. In some cases, these forces vary widely as task conditions change. The control device is being affected by the forces external to the vehicle such as tire friction from ground contact. In this case, active or dynamic force feedback results from the interaction of the vehicle with the environment as a normal component of the real-world vehicle operation and becomes an integral part of the vehicle operator's perception of vehicle control. For example, steering wheels on an automobile are connected to a steering shaft and to steering gear on the front wheel. The wheels, in turn, are subject to tire friction or resistance, which is fed back to the driver through the steering wheel itself. The additional forces required to overcome this resistance are variable or dynamic in that they change with conditions.

Dynamic force and force feedback perception are even more critical to aircraft flight control. In aircraft, control forces required to displace the control device, such as a control wheel or stick, vary dramatically with the speed of the aircraft. This is due to the fact that the control surfaces of an aircraft to which a control device is connected become increasingly effective as the speed of the aircraft increases. This is due to the resultant increase of airflow over control surfaces. Due to these changes in control effectiveness with airspeed, pilots must adjust the forces applied to controls with the changes in airspeed.

In both aircraft and ground vehicles, force perception is especially important during critical maneuvers. For ground vehicles, these are typically maneuvering tasks such as driving on curves, hazard avoidance, and off-road driving. For aircraft, accurate force perception is particular critical at low air speeds such as during takeoff and landing. In both cases, environmental components are added to the control force perception problem. Wind gusts and rapid changes in wind speed or direction are particularly problematic for vehicle control force perception.

Basic Force Perception

In order to design simulated control device forces that can replicate static and dynamic force perception, the designer requires an understanding of how well humans can discriminate forces when applied to objects such as control devices. Determining this discriminative ability provides the vehicle simulator designer with guidance as to how much control force is needed, under what conditions, and within what tolerances. Recreation of real-world vehicle control forces in the simulator to such a degree that the operator is unable to discriminate the difference is a means of achieving high perceptual fidelity.

In basic force perception, the biomechanics of fingers, hand, and foot movements play an important role. As a rule, the greater the number and size of muscle tissue used to exert forces over a control device, the lower the discriminative ability of the user will be when perceiving differences in the forces required to move the device. Thus, fine hand movements used to control a device such as a stylus or a small switch or a knob are much more sensitive to small force changes than the movement of the foot control device such as an accelerator or brake pedal. It is axiomatic that large muscle groups such as those in the legs and arms are capable of exerting forces much larger than those of the fingers or hand. Thus, upper force levels need to be kept in mind for controls simulation to assure that the device can withstand the high levels of the forces used. For example, in an emergency, braking forces applied to a brake pedal can easily reach 400 N (90 lb).

Finger Force Perception

Due to the rise in interest in virtual environments in the computer industry, more research attention has been paid to the perceptual processes that are involved in force perception. This is particularly true for finger and hand force perception as this supports the design of more perceptually realistic control devices in virtual environments.

Among the simplest means of assessing finger force perception is to measure the individual's ability to discriminate the forces required in the thumb–index finger grasping movement. In studying this movement, hand, wrist, and forearm muscle group are systematically restricted so that each muscle group involved in the grasping act can be isolated. Table 4.1 derived from the study by Tan et al. (1994) reveals the maximum forces that can be produced at each joint segment.

The study was conducted using a small sample of two males and one female to measure the maximal control of the force on a load cell for 5 s. Though a small

TABLE 4.1
Average Maximum Controllable Force

	PIP	MCP	Wrist	Elbow
Mean	36.4 N	35.1 N	51.8 N	75.2 N
SD	2.7 N	1.5 N	1.9 N	1.6 N

sample was used, the study demonstrates clearly how each of the potential muscle-joint segments is involved in the grasping response. (Note that the principal muscle groups for finger flexion are in the palm of the hand and forearm, not the finger itself). The finger joints proximal interphalangal (PIP) produced an average maximum force of 36.4 N (8.2 lb) and the metacarpal phalangal (MCP) a force of 35.1 N (7.9 lb). However, dramatically increased force is available if the wrist is allowed to move. In this case, the maximum force increases to 47.7 N (10.7 lb). With the elbow free to move, the maximum force increases to 68.6 N (15.4 lb) and up to 85.3 N (19.2 lb) when the shoulder was allowed to move. The authors also examined the accuracy with which the force can be produced. For the PIP, MCP, wrist, and elbow, the accuracies were 1.89%, 1.98%, 1.08%, and 1.28% of the required force production.

A study by Pang et al. (1991) examined force perception in thumb–forefinger grasping. Using a horizontal track, subjects were required to grasp two plates attached to the track using the thumb and forefinger applying a squeezing force to the two plates. A varying resistance force was then applied to the two plates and subjects had to compare this force to a reference force. The average JND for the task was 7%, subjects being consistently able to discriminate force differences using a reference force of 5.0 N (1.1 lb). The JND was consistent across forces ranging from 2.5 N (0.6 lb) to 10.0 N (2.2 lb).

For the vehicle simulator designer, the positioning of control devices within the architecture of the vehicle cockpit is an important element if perceptual fidelity is to be achieved. The positioning of the control devices and whether the user can exercise the same muscle groups to operate those controls is important in achieving biomechanical forces comparable to those in the real vehicle. Allowing the operator to apply excessive forces or to restrict the operator to subnormal forces both create issues for perceptual fidelity. Avoiding the introduction of abnormal force perceptions is essential for the experienced operator who will be forced to alter their control behavior to accommodate to the simulator. Inaccurate force requirements also introduce problems for novice operators who will not be prepared for the control forces required of the real vehicle operation.

Hand Movement Force Perception

Perhaps more germane to the issue of vehicle simulator design, especially control force simulation, is the ability to discriminate forces in hand movements. Most hand control devices in vehicles allow some hand movement and, in some cases, forearm and shoulder movement as well. The nature of the hand movement is important to achieve a high degree of perceptual fidelity as the discrimination of force required to move hand control devices will change depending on the restrictions placed on hand movement in the real vehicle.

In the case of hand movements of simple devices such as a handheld stylus, relatively small forces are required for movement. In a study by Yang et al. (2000), force magnitude perception for the movement of a computer stylus was measured in different directions along the lateral and longitudinal axes. Only isolated hand movement was allowed and subjects were exposed to varying levels of forces in 0.2 N (0.04 lb) increments relative to a reference force of 1.5 N (0.34 lb) Disturbances to the stylus were applied in different directions (0°, 45°, 90°, 135°, or 180°) from the axis of

TABLE 4.2

Force Perception of Disturbances to the Control of a Computer Stylus

	Force Direction				
	0°	45°	90°	135°	180°
Force	0.47 N	0.92 N	0.82 N	0.65 N	0.52 N
JND	31%	61%	54.5%	43%	34.5%

movement relative to the reference force. The subjects were required to move the stylus in reference to a visual target, which moved horizontally on a computer screen in front of them. The results in Table 4.2 are for the average of two hand movement speeds tested. The perception of force in the operation of the stylus varies significantly depending on the direction of the force relative to the movement of the stylus.

The study found that the ability of the subject to discriminate the disturbance forces on the control device varied with the relative force direction of the disturbance to the control. Force discrimination (JND) in the same axis as hand movement (force direction of 0° and 180°) was 31% and 34.5%, respectively, or nearly one-third of the reference force needed in order to discriminate a force difference. Thus, forces that are *congruent* with the direction of hand movement of the control resulted in the most accurate perception of force by the subjects. The authors note that this is some three times larger than the force discrimination levels found in previous studies when no movement of the hand was allowed. Thus, force perception for the stylus control is dependent on whether or not the hand is in motion as well as the direction of the force relative to that motion.

Steering Wheel Force Perception

Force discrimination of a small stylus movement, although relevant to similarly sized control devices, is not likely to be the same as the discrimination in larger control devices such as steering wheels or aircraft control wheels.

For automobiles and trucks, the conventional steering wheel is still the primary control device for lateral control in ground vehicles. In both vehicle categories, the addition of power-assisted technology has significantly improved the ease by which the vehicle operator can maintain control over the vehicle. Early introduction of power-assisted steering often met with driver disapproval because it resulted in excessively light control forces at higher speeds. This level of force feedback is contrary to the findings that drivers want an *increase* in torque feedback as speed increases (Bertollini and Hogan, 1999). Variable speed, power-assisted steering evolved over the decades to improve the perceived steering wheel *feel* by increasing control force feedback from the road surface. This desire for increased steering wheel torque at higher speeds may be related to a general feeling or the need for an increased positive control over the vehicle path. At high speeds, small steering wheel deviations have dramatic effects on the vehicle path and the desire for stronger force feedback may have been a desire to avoid dangerous steering wheel inputs at high speeds.

This need for a greater force feedback with increasing speed needs to be considered as important to the perceptual fidelity of steering wheel force feedback.

Driver perception of steering wheel characteristics can be divided into two components: (1) torque and (2) force feedback. The former, torque, is generally measured by the distance between the hands and the steering wheel column. Other things are equal, the greater the distance the hands are from the column the less torque feedback will occur at a given level of force applied to the steering wheel shaft. Force feedback for most routine driving tasks is between 0.1 Nm (0.07 ft-lb) and 5 Nm (3.7 ft-lb). The force feedback to the driver has been experimentally isolated from the torque and shown to be the more important feedback component (Newberry et al., 2007), although both torque and force contribute significantly to control behavior.

A number of studies have attempted to determine the effects of steering wheel feedback (measured as torque) on driving tasks typically in passenger car driving simulators. In a lane change task, increases in the torque *gradient* applied to the steering wheel (0, 9.0 Nm/rad [0.12 ft-lb/deg], and 17.9 Nm/rad [0.23 ft-lb/deg]) significantly decreased the variance in steering wheel angle (Liu and Chang, 1995). These investigators also examined the effects of variations in steering wheel torque feedback. Torque forces ranged between 2.8 Nm (2.07 ft-lb) and 5.67 Nm (4.2 ft-lb). Comparing torque feedback applied to the steering wheel column to a no torque feedback condition, they found no effect on steering wheel angle variance in simulated curve driving. A further increase of torque feedback to a force of 8 Nm (5.9 ft-lb) also had no effect. However, in a skid recovery maneuver from a sharp curve, torque feedback significantly improved the recovery even with the lowest torque level available (2.8 Nm or 2.1 ft-lb).

Another study conducted by Toffin et al. (2003) used a driving simulator equipped with a motion platform. This study also failed to find any effects of torque on steering wheel angle variance when the torque was varied between 1.5 (1.1 ft-lb) and 2.5 Nm (1.8 ft-lb) on curved roads. However, a zero torque feedback made driving very difficult, if not impossible. In a second experiment, the authors compared linear torque feedback with nonlinear (parabolic) feedback. The nonlinear feedback profile resulted in little or no torque available within the steering wheel angle was within ±20° of neutral, while increasing to 4 Nm (3 ft-lb) for steering wheel angles beyond these values. No effects were found on either the steering wheel angle or on the lateral acceleration of the vehicle when either linear or nonlinear torque gradients were compared.

The effectiveness of different technologies for providing torque feedback was assessed by Mourant and Sadhu (2002) using a fixed-base simulator. Two low-cost, PC-based steering wheels using either a spring-based or a torque motor were compared. Driving performance was compared on curved roads of either 100, 200, or 300 m in radius. The higher torque levels available to the torque motors resulted in drivers hugging the left hand of the road in curves more often than when using the lower torque spring-based system. Variance of lane position was also reduced with the torque motor system in the more demanding curved road of 100 m radius.

The role of steering wheel torque feedback is heavily influenced by the task required of the driver. Specifically, the more demanding tasks, such as control in tight road curves and skid recovery, are more likely to benefit from the availability of torque feedback when compared to less demanding lane maintenance tasks on straight roads. For typical automobile simulators, torque levels of 3 Nm appear adequate for most driving tasks. Schumann (1993) cited in Liu and Chang (1995) provides psychophysical data indicating a lack of driver sensitivity to torque levels less than 2 Nm (1.5 ft-lb), whereas torque levels above 5 Nm (3.7 ft-lb) produced excessive levels of driver fatigue.

The principle utility of steering wheel feedback in vehicle driving appears to be in preventing overcontrol, particularly under more demanding conditions that require precise steering wheel inputs. In this regard, it shares some features of physical motion in driving in which this type of cue is effective but only in more demanding tasks such as high-speed curves or in the presence of disturbance inputs.

Achieving acceptable levels of perceptual fidelity with regard to steering wheel torque feedback requires that the levels are in the range of 3 Nm (2.2 ft-lb) to 4 Nm (3 ft-lb). This applies to normal or routine driving tasks, including curve to straight transition tasks and passing maneuvers. Extreme driving characteristics such as those found in Formula 1 racing may require steering torque levels much higher. This may also be the case with certain off-road vehicle operations. The precise torque values for these nonroutine driving tasks need to be determined empirically.

The biomechanical architecture of the steering wheel placement within the simulator also needs to be considered. As the torque is a function of where the driver's hands are in relation to the steering wheel column, replacing a standard-sized steering wheel with a smaller one is acceptable provided the same level of torque feedback is maintained. This means that torque levels will be greater at the steering wheel rim if the steering wheel diameter is reduced, other things being equal. Conversely, replacing a small diameter steering wheel with a larger steering wheel in an attempt to add physical realism will effectively reduce torque feedback unless the device's torque amplitude can be increased. Sacrificing perceptual fidelity to achieve the level of increased subjective realism will result in less perceptual fidelity and less realistic control behavior as well.

Aircraft

Thus far, only force feedback affecting automobile steering wheel inputs has been discussed. The two main hand control devices used in aircraft are the control wheel and the control stick. The former is a partial steering wheel device mounted on the control shaft that can be rotated or moved in and out. Unlike automobiles and trucks, the control wheel does not require continuous input in order to initiate a turn. The control wheel in an aircraft is a *rate* control device in which a given control-input initiates a roll maneuver. The control wheel is returned to the neutral position when the desired turn rate is achieved. The turn itself is maintained by aerodynamic forces acting on the aircraft. To end the turn, the pilot cancels the rate change by moving the control wheel in the opposite direction.

Control wheel inputs to change aircraft pitch angle requires the pilot to maintain the control wheel at the position necessary to achieve a given aircraft pitch angle (up or down). Unlike the control behavior of turns using the rotation of the control wheel, the control behavior for pitch control requires the pilot to push forward or pull backward. As the aircraft will lose airspeed when the pitch angle is increased, control forces needed to maintain the pitch angle need to be increased (unless power is added to increase airspeed). As maintaining a fixed pitch angle continuously can be fatiguing for a pilot; aircraft are equipped with pitch trim controls to allow the pilot to reduce the forces required to maintain a continuous pitch angle.

The control stick, commonly found in smaller aircraft, including fighter aircraft, operates in the same manner as the control yoke except that the control stick is mounted on the floor forward of and in the medial plane of the pilot's body. This position means that the biomechanical forces differ from that of the control wheel. On account of its location, the control stick movements can rely not only on hand and forearm musculature, but on the upper arm and shoulder muscles when heavy forces for lateral control of the stick are required. In contrast, control yoke movements needed to initiate the same roll motion in the aircraft are necessarily more constrained because they are fixed in rotation about a central control column.

As with automobiles and other ground vehicles, aircraft control systems have evolved over the years from simple pulley and cable systems to hydraulically boosted controls, and finally to the computerized fly-by-wire systems of today. Each of these systems presents their own unique problems for perceptual fidelity and force feedback. For the vehicle simulator designer, the problem remains the same as to how to recreate the control forces that are perceptually indistinguishable from the real aircraft.

The question then, as with ground vehicles, is how much of a difference between the real and the simulated vehicle control force feedback must exist before a pilot can perceive the difference. Second, what effect will the reduced or nonexistent force feedback have on how the pilot controls the simulated aircraft?

The similarity between the steering wheel of a ground vehicle and rotation of the aircraft control wheel suggests that the force discrimination data found in automobile steering wheel control behavior may be applicable to aircraft control wheels. This would likely be the case if the range of forces applied to the aircraft control wheel is comparable to that of an automobile. Similarly, the fore-aft aircraft control wheel movement discriminatory ability in some aircraft is likely to be comparable to the fore-aft hand movement found in the study by Yang et al. (2000). In this study, JND was found to be 31% of reference force of 1.5 N (0.34 ft-lb). In this case, users of the aircraft simulator control yoke movement forces at this low level are more likely to exist for light aircraft and in-flight nonaerobatic maneuvers. Normally, only very light forces are required for flight control in such tasks.

There are surprisingly little objective data on pilot perception of control wheel or control stick force perception. Flight simulators are usually not assessed for objective force perception. Instead, their handling characteristics are evaluated by experienced test pilots who determine whether or not the flight simulator control feel is comparable to the real aircraft. Subjective rating scale data might be used, such as the Cooper–Harper rating scale, but objective force perception data are lacking. Instead, a considerable effort has been expended to model pilot-control behavior as an optimal control system design. These models have concentrated on quantifying purely sensory component functions and not on force perception. Indeed, the unified theory of aircraft handling qualities proposed by Hess (1997) explicitly eliminates consideration of control and force feel perception of the pilot. This apparently is due to the need to eliminate the uncertainty associated with this aspect of pilot behavior.

Apart from the inherent problems of subjective ratings as well as the use of non-representative users, there is little evidence to determine what aspect of the control system force feel matters or whether it really matters in aircraft simulation. Again, subjective realism is not, in itself, an adequate means of evaluating simulator perceptual fidelity as defined in this text. Evidence is therefore needed that variations in control force feel have at least a measurable effect on pilot behavior in order to determine whether investment in force feedback technology is worth the cost. In the absence of such data, we are reliant on subjective opinion. It may be the case that complex control force feedback systems may not be required for control devices used in more routine tasks. This is the case with automobile steering wheel input and may well be the case for aircraft control devices as well. If so, the simulator user community may need to only invest in more sophisticated and costly force feedback systems in advanced driver or pilot training in which the task demand is much higher. This would reduce the cost of simulator control force feedback simulation for use in novice training or in less demanding task conditions.

One of the few studies examining control force feedback provides some insight into the effects of control force feedback on behavior (Cardullo et al., 2011). The study examined the role of control force feedback in a compensatory tracking task using simplified plant dynamics. The task of the pilot was to nullify disturbances in roll under two levels of plant dynamics and two levels of force feedback fidelity. Two groups of pilots were trained to asymptote: one group using the low fidelity force feedback and the other using the high-fidelity force feedback. High fidelity was determined by the use of control force feedback parameters, which corresponded to those with design features found in the literature to yield *good performance* from operators. A side-stick controller was then designed to incorporate the following control dynamics: A force gradient of 45.2 N per cm (4 lb/in.), damping ratio of 0.7, and a breakout force of 6.7 N (1.5 lb).

The low fidelity controller was a consumer-level control stick commonly used in personal computers for flight simulator gaming. It had a force gradient of only 7.9 N/cm (0.7 lb/in.), a damping ratio of 0.135, and a breakout force of 0 N. These device control forces differ dramatically from the characteristics of the high fidelity control described earlier.

The authors of this study developed a control device based on what previous research had determined to be a *good performance*. This behavioral approach to the design of a controller is analogous to a behavioral measure of perceptual fidelity. However, this study did not employ the dynamics of a real vehicle. Rather, it used dynamics that generically represent low and high levels of control difficulty as found in previous studies. Thus, direct translations of the results of this study to specific vehicle simulations are difficult to make.

However, the study results may shed light on how variations in control dynamics affect pilot behavior. The assumption being that this is a valid indicator of how pilots behave in controlling vehicles. Indeed, although the study found no significant difference between the two levels of control device fidelity in the case of the task of low difficulty, it did find significant differences in performance when tracking with difficult plant dynamics. Notably, the plant dynamics in the more difficult task require the pilot to respond much more quickly to a disturbance. If the pilot is slow to respond, the target would quickly accelerate away. In contrast, the low difficulty task allowed more time to null the disturbance input as the disturbances had the effect of moving the cursor with a constant velocity. One or more of the three parameters of physical control force feel that gave the advantage to the higher fidelity device in the difficult control task appears to have been eliminated when the pilots performed the simpler task. One of these, the force gradient, is known to affect control behavior in aircraft-handling qualities literature and is likely to play a major role in the perception of control force. Pursuing these differences in the effects of force gradients (and other attributes) more systematically using more realistic vehicle dynamics would be of great help in establishing design guidelines for simulator control systems.

Cardullo et al. (2011) also addressed what would happen when those trained with the low fidelity control device would transfer to a high fidelity device. Again, the study found no differences between low and high fidelity devices in the degree to which skill training is transferred for the low difficulty task. However, there was a significant difference in transfer between the two fidelity levels when the high difficulty task performance was examined. Analyses of the control behavior of the two groups suggest that the high fidelity group is developing a control strategy that can more readily transfer to the highly difficult task and that consequently these pilots perform much better during transfer than those trained with low fidelity device. Finally, the low- and high-fidelity transfer group eventually performed similarly after many trials. Pilots appear to have developed a control strategy with the low fidelity control device that allowed them to perform at a level comparable to those using the high fidelity control device. This control strategy development, however, took time to achieve.

Although not directly relevant to specific aircraft vehicle dynamics, the study is suggestive that the impact of perceptual fidelity in control behavior is likely to depend on how demanding the vehicle control task happens to be. Aircraft with particularly demanding vehicle dynamics such as helicopters are likely to benefit more from control devices with high perceptual fidelity force feedback than those aircraft that have high inherent stability. However, it is possible that these more stable aircraft may also encounter unusual control task conditions, which might be comparable in difficulty to those that a helicopter pilot might encounter regularly. Such control conditions can be produced by extreme weather and wind events as well as equipment failures and malfunctions.

More studies are needed to examine the role of control device force feel on pilot behavior in a more realistic vehicle and task conditions before determining the true value of aircraft control force feel fidelity. Thus far, the findings for control force feel in vehicle operating perceptual-motor behavior appear to mimic those found in physical motion. That is, its influence on behavior is more likely to be revealed under more demanding control conditions rather than under more routine control conditions.

In any case, the requirement for sophisticated force feedback is probably not justified for routine maneuvers in which control force requirements are modest and the vehicle dynamics are relatively stable. In this case, less sophisticated force feedback systems that provide modest force feedback levels that limit excessive control inputs may be enough. These force feedback systems are more likely to be characterized by a constant force feedback, which can be delivered by simple torque motor systems.

Thus, even relatively simple, PC-based, force feedback control devices such as torque motor systems may be sufficient for many aircraft simulator applications.

Although there is sufficient objective and subjective data to define at least a preliminary level of perceptual fidelity in automobiles, more data are needed to support the conclusions above for aircraft control force feel fidelity. This deficit in objective data is likely due to the assumption in flight simulator design that force feel design issues are a source of unwanted noise in pilot-vehicle control modeling. In addition, the industry tends to rely on subjective evaluations for force feel by the test pilot population and then only as an integral part of the pilot-vehicle control system, not as isolated components.

VEHICLE FOOT CONTROLS

Vehicles require some form of manual control device to operate the vehicle. For ground vehicles, such as automobiles, the brake and accelerator are the most common along with the clutch pedal in the manual transmission vehicles. Aircraft foot controls include rudder pedals, which may also incorporate toe brakes. Once again, these controls are operated without direct visual input as to their displacement. The operator must rely solely on proprioceptive and tactile feedback from the foot and other parts of the leg to determine the correct pressure to apply.

Automobile Foot Controls

The amount of pedal pressure an operator must apply is an important issue for all automobile and truck manufacturers. Excessive requirements for pedal pressure will result in operator control difficulties as will the need to have very little pressure. The latter can result in vehicle over control such as an excessive acceleration or braking. Brake pedal pressure is a particular problem because it represents a safety critical control device at least as important as a steering wheel control. Modern vehicles typically include power assist breaks, so that high brake pedal pressure is not required. A recent naturalistic study of brake pedal force used by drivers in automobiles revealed a range of 4–30 N (0.9–6.7 lb) between the 5th percentile males and the 95th percentile females (Gkikas et al., 2009). Average pedal pressure was 78.6 N/cm^2 (114 psi) with a standard deviation of 65.6 N/cm^2 (95.1 psi).

This study also revealed dramatic gender differences in the amount of the applied braking pressure. Average brake pressure was four times as high for the 75th percentile females as it was for the 5th percentile male. Although the overall braking pressure range was about the same for each gender, female braking averaged nearly twice that of males.

For vehicle simulator design, the biomechanics of foot pedal placement within the architecture of the vehicle cockpit will play a significant role in how the brake pedal pressure is applied. A recent study of foot pedal pressure discrimination revealed that discrimination could be accurately modeled mathematically provided optimal conditions were met (Tanaka et al., 2009). At foot angles of 115° (25° past vertical), 45° of knee flexion, seat pan at 250 mm (9.8 in.) above the floor, and seat tilted slightly back at 15° past vertical, the ability of drivers to discriminate foot pedal pressure could be predicted with a high degree of reliability ($R^2 = 0.87$).

Force perception at the brake pedal was estimated by subjects beginning with 20 N (4.5 lb) as a reference. The subject was then exposed to a comparison foot pedal force stimulus ranging from 5 N (1.1 lb) to 40 N (9.0 lb). The subject would then express an estimate as to the proportion of force of the comparison stimulus to the standard force. Regression analysis was then performed on the resulting dataset. The formula describing this relationship between the perceived force (F_p) and comparison force (F_r) was found to be

$$F_p = 19.273\, L_n(F_r) + 11.76$$

Thus, the perceived foot pedal force was found to be approximately proportional to the logarithm of the comparison force magnitude examined in this study. However, the change in the relationship between the perceived force and actual force was dramatic when the foot angle was changed from 115° to 145° (55° past vertical). The following formula describes the new relationship:

$$F_p = 34.483\, L_n(F_r) - 36.357$$

The discriminative function of the subject is changed significantly at this pedal configuration. The constant applied to the logarithm of the force is nearly double that of the 115° pedal angle.

Aircraft Foot Controls

Aircraft foot control force discrimination has similar characteristics to ground vehicles in that they are heavily influenced by the biomechanics of the foot joint, knee joint, and associated muscle groups.

As with passenger and truck manufacturing, manufactures of aircraft must consider the ergonomics of the foot control design to assure optimal configuration of seat height from the cockpit floor, pedal angle, and seat pan distance from the pedal controls. In a study of 100 pilots by Hertzburg and Burke (1971), pedal angle was varied between 5° of foot flexion to 55° of foot extension. These two values covered the extremes of allowable pilot foot angles in the study. The study found that maximum foot forces can only be achieved within a range of 15° to 30° past vertical with peak pedal forces achieved within 20° to 30° passed vertical—a pedal angle range of only 10°.

This study examined the foot angle as it applies to the use of toe brakes on aircraft rudder pedals. For most aircraft, braking controls are integrated into the rudder pedals as toe brakes. The forward rotation of the foot is necessary to apply the toe brakes. This foot rotation is not employed in the use of rudder pedals. Instead, the pilot uses the entire force of the leg to *push* the pedal forward. As foot angle rotation occurs in the accelerator pedal controls for most ground vehicles, these data might also be applied to the operation of passenger car and truck pedal design as well.

Actual discriminative forces for foot pedal pressure on aircraft rudder pedals are lacking, although parameters such as the maximum force required for pedal displacement (300 N or 67.4 lb) and maximum excursion degrees for full on–off operation (30°) can be found in published guidelines used by the aircraft manufactures.

TABLE 4.3

Airbus A 300–600 Rudder Pedal Forces and Displacements as a Function of Airspeed

Airspeed	Max Pedal Force	Max Travel
165 kts	294.0 N	10.2 cm
240 kts	157.0 N	3.7 cm
310 kts	127.0 N	1.7 cm

An example of these types of data on pedal force displacement of an airliner is shown in Table 4.3. The table shows the relationship between airspeed and the forces required for displacement at each speed. Rudder pedal forces in aircraft, as with other controls, are dependent on airspeed. The faster the speed, the less force will be required to effect the positive control.

In the study that reported these values (Tan and Hernandez, 2004), an attempt was made to match the physical values of the pedal force and the travel for this aircraft and an aircraft simulator. Although the values of the simulator pedal force and a maximum pedal travel closely matched that of the aircraft, the matches were not identical. Although this strategy of physically replicating certain attributes of a control seems logical, the inevitable differences in control forces between the simulator and the aircraft could result in behaviors that are not comparable for the two systems. Short of perfect replication, the designer cannot be certain of what the impact of even small differences in physical fidelity will be on behavior. In order to achieve the desired behavior, it is necessary to achieve perceptual, not simply physical, fidelity. In this case, the differences between the simulator and aircraft in maximum pedal force and maximum pedal travel may or may not have been sufficiently small as to be beyond the level of pilot's discriminative ability. A decision regarding whether or not perceptual fidelity had been reached in the simulator can only be made if one knows the ability of the pilot to discriminate one level of pedal force (or travel) from another. In this case, the JND for force or displacement of a foot pedal could be used as a guide to keep the differences between the simulator and the aircraft to levels that would not be perceptible. It would be most desirable if JNDs for aircraft pedal force and displacement perception were available, but a comparable vehicle system might be used provided the forces and displacement level ranges are similar to the aircraft. JND values for clutch pedals found by Tanaka et al. (2009) could be used and adjusted higher if the designer wishes a more conservative estimate. The residual $(1 - R^2)$ of the coefficient of determination in this study was 0.13. This can be used to estimate the error variance associated with the JNDs for force perception. However, the pedal forces in this study ranged from 25 to 200 N (5.6–45 lb), which is somewhat below the range of pedal forces for this large aircraft.

The data from Southall (1985) may be more appropriate in this case. Southall evaluated pedal force discrimination in the use of clutch pedals in trucks. The range of pedal forces for this study was between 89 and 445 N, much closer to that of the Airbus 300–600. The JND for these pedal force levels produced a consistent

Weber fraction of 0.07 (7%). Weber fraction means that the JND remained constant across the range of pedal force values in the study. As a given JND is the midpoint of the distribution of differences, at least half of the pilots might have perceived a difference in maximum pedal force between the simulator and the real aircraft in this study. In actuality, a more conservative JND would be used one, for example, that would set the detection of differences to a much lower statistical level, say the 5th percentile of the JND distribution. It is not known whether these perceptual differences between the pedal forces in the aircraft and the simulator resulted in pilot performance differences between those of the simulator and the aircraft, but reliance on perceptual rather than physical fidelity would probably have been more desirable.

It might be argued that the use of JNDs or Weber fractions derived from laboratory studies is not necessarily applicable to the design of virtual environments due to the high task loading of users in these environments. Pilots operating in simulated aircraft environments are attending to many other tasks than simply the perception of pedal force, for example. However, the use of these data by designers should consider that the consequences of erring on the side of over estimating the perceptual ability of their users is preferable to that of designing a virtual environment that has low perceptual fidelity with all its attendant implications for the utility of that environment.

SUMMARY

Perceptual fidelity in the design of virtual environments must include considering how manual control of components of the system are being affecting by the simulation of force feel and feedback of operator controls. This is especially true in the design of vehicle simulators in which the controls are often being operated without the aid of vision. Control forces, as we have seen, can be measured physically without any reference to human perception. However, matching these forces exactly in the simulator is a very difficult task and is often very expensive to accomplish. In any case, physical replication that is less than exact leaves the designer with doubts as to whether the design will be close enough such that the remaining differences would not be noticed by the end user. As perfect replication of control characteristics is lacking, the designer then relies on evaluation by expert operators. Indeed, it has been the common practice in the design of vehicle simulators to evaluate force feel and feedback simulation by using expert operators who assess the fidelity subjectively. The process is often followed by a period of *tuning* in which control forces are varied in order to attempt some degree of at least subjective realism. An understanding of the basic biomechanics of manual control and of the human perception of force feel and feedback provides a more orderly and rigorous approach to the design and evaluation process. Particularly important will be an understanding of which elements of the control device characteristics (force gradient, damping ratio, and breakout force) are affecting what aspect of the control behavior. The values used in the control design by Cardullo et al. (2011) may be helpful in isolating some of the key design characteristics. Systematic variations of these characteristics in a vehicle simulator and an examination of their effects on behavior may lead to better means by which high levels of control feel and feedback can be achieved in virtual environments.

The use of JNDs and other measures of human control force perception will continue to play a role in the design of virtual environment control components. The use of research simulator testing will also remain an important means of determining the perceptual fidelity of control force feel and feedback.

REFERENCES

Bertollini, G.P. and Hogan, R.M. 1999. *Applying Driving Simulation to Quantify Steering Effort Preference as a Function of Vehicle Speed*. Warrendale, PA: Society of Auto Engineering, 1991-01-0394.

Cardullo, F.M., Stanco, A.A., Kelly, L., Houck, J., and Grube, R. 2011. A transfer of training study of control loader dynamics. *AIAA Modeling and Simulation Technologies Conference*, August 8–11, Portland, OR.

Gkikas, N., Richardson, J.H., and Hill, J.R. 2009. A 50–driver naturalistic braking study: Overview and first results. In P.D. Bust (Ed.), *Contemporary Ergonomics* (pp. 423–431). London, UK: Taylor & Francis Group.

Hertzburg, H.T.E. and Burke, F.E. 1971. Foot forces exerted at various brake-pedal angles. *Human Factors*, 13, 445–456.

Hess, R.A. 1997. Unified theory for aircraft handling qualities and adverse aircraft-pilot coupling. *Journal of Guidance, Control, and Dynamics*, 20, 1141–1148.

Liu, A. and Chang, S. 1995. Force feedback in stationary driving simulation. *Proceedings of the IEEE Conference on Systems, Man, and Cybernetics*, October 22–25, vol. 2, 1711–1716. IEEE.

Mourant, R.R. and Sadhu, P. 2002. Evaluation of force feedback steering in a fixed based driving simulator. *Proceedings of the Human Factors and Ergonomics Society 46th Annual Meeting*, 2202–2205. Los Angeles, CA: SAGE Publications.

Newberry, A.C., Griffin, M.J., and Dowson, M. 2007. Driver perception of steering feel. *Journal of Automotive Engineering*, 221, 405–415.

Pang, X.D., Tan, H.Z., and Durlach, N.I. 1991. Manual discrimination of force using active finger motion. *Perception and Psychophysics*, 49, 531–540.

Schumann, J. 1993. On the use of proprioceptive-tactile warning signals during manual control—The steering wheel as an active control device, PhD Dissertation. Universitat der Bundeswehr Munchen, Neubiberg, Germany.

Tan, D. and Hernanadez, E. 2004. Use of the Vertical Motion Simulator in Support of the American Airlines Flight 587 Accident Investigatio. *AIAA Modeling and Simulation Technologies Conference and Exhibit*, August 16–19, Providence, Rhode Island.

Tanaka, Y., Kaneyuki, H., Tsuji, T., Miyazaki, T., Nishikawa, K., and Nouzawa, T. 2009. Mechanical and perceptual analyses of human foot movements in pedal operation. *Proceedings of the 2009 IEEE International Conference on Systems, Man, and Cybernetics*, San Antonio, TX.

Toffin, D., Reymond, G., Kemeny, A., and Droulez, J. 2003. Influence of steering wheel torque feedback in a dynamic driving simulator. *Proceedings of the Driving Simulation Conference: North America*, October 8–20, Dearborn, MI.

Southall, D. 1985. The discrimination of clutch-pedal resistances. *Ergonomics*, 28, 1311–1317.

Yang, X.-D., Bischof, W.F., and Boulanger, P. 2000. Perception of haptic force magnitude during hand movements. *Proceedings of the IEEE Conference on Robotics and Automation*, May 19–23, Pasadena, CA.

Willingham, D.B. 1998. A neurological theory of motor skill learning. *Psychological Review*, 105, 558–584.

5 Multimodal Perception and Vehicle Control

INTRODUCTION

In the preceding chapters, many of the perceptual issues central to the issue of perceptual fidelity in vehicle simulation were discussed. Assessments of the sensitivity of human sensory and perceptual processes for individual perceptual systems such as vision were included. The basic thresholds to various individual stimuli were noted as well as the ability to discriminate the relative intensity of stimuli. Finally, the studies examining the influence of these stimuli on the performance of tasks in simulated and real-world vehicle operations were reviewed.

The operator of a vehicle in the real world, however, is often experiencing stimulation in more than one modality at the same time. Two or more modalities, such as visual and vestibular, are combined in some tasks to the extent that each contributes in supporting task goals. Some of this effect of multimodal perception was seen earlier in the changes that occurred when vestibular stimuli were accompanied by visual stimuli (the oculogyral illusion). The result was a dramatic improvement in the ability to detect physical motion. Examples were also given in which the combined visual–vestibular cues affected the control behavior differently than when either cue was presented alone. This interactivity of multimodal cues complicates the decision for designers who may wish to eliminate one or more modality from a design to save cost.

MULTISENSORY SUBSYSTEMS OF POSTURAL STABILITY AND LOCOMOTION

The reason for these interactions among sensory systems becomes more obvious when one examines the various sensory subsystems within the human nervous system that the sensory modalities subserve. When discussing the importance of the human visual system for tasks such as the perception of object movement, it is often forgotten that vision as well as vestibular and somatosensory systems evolved to support proper postural stability and locomotion.

Some of these multisensory interactions occur at relatively low levels of the central nervous system. For this reason, they are often not subject to conscious awareness. The responses to changes in posture either voluntary or due to a disturbance are typically carried out automatically with little conscious attention paid to the details. These multimodal subsystems have evolved over the millennia to aid in survival. Thus, all animals need to maintain control of their posture to function properly. Moreover, the animal needs to maintain balance in relation to the earth vertical. That is, there is a natural tendency to maintain an upright posture in order to be prepared to respond quickly to events. The failure to prepare for a rapid response often means the loss of prey or death from a predator.

Postural stability mechanisms are ingrained in the nervous system of all mammals, including humans. The postural stability control system mirrors that of the closed-loop control system in vehicles. In place of the control-input device of a vehicle, the human musculature system is engaged in this balance control system to orient and stabilize the body in response to changes, either initiated voluntarily or as a result of external disturbances to body position.

Similar mechanisms are at work when humans move through the environment. Locomotion is essential to survival for all animals and the multisensory subsystems that support postural stability also supports locomotion. The perception of self-motion and the heading or movement trajectory of that motion that support locomotion are extracted from optic flow patterns in the visual field. Again, humans and other mammals have evolved in a structured world in which the perception of self-motion has been extracted from the regularity of optic flow patterns that occur with the movement of the head. Nonvisual cues also accompany such movement. During longitudinal acceleration and deceleration, vestibular inputs are received by the subsystem of locomotion. In addition, somatosensory inputs from the movement of joints and musculature as well as tactile feedback from skin mechanoreceptors enhance the perception of self-motion. As with postural stability, control of velocity and trajectory once again is dependent on musculature signal processing. Again, movement control of body mimics the movement control of a vehicle. The vehicle movement control is, however, modified by the vehicle's dynamics in the same manner in which body dynamics modify the body's response to inputs.

Treating multimodal cues as subservient to a particular perceptual subsystem, such as postural control or locomotion, rather than just as a collection of cues reduces the risk of creating virtual environments which could produce unrealistic or undesirable behaviors in their users. These include behaviors that are temporary adaptations to an artificial perceptual environment, which may be wholly unrelated to that found in real life. The failure to understand perceptual cues as supportive of more complex subsystems can result in debilitating phenomena such as simulator sickness. The latter, a form of motion sickness, has plagued the development of virtual environments. Simulator sickness is a result of the failure of designers to understand the complex, multimodal interactions that support many aspects of human performance. Avoiding these phenomena should be a priority for the designers of virtual environments.

THE MULTIMODAL PERCEPTION SYSTEMS

The neuroanatomy of sensory systems such as postural stability are often revealing of the potential interaction among the different perceptual modalities. Sensory systems that share common components such as the visual–vestibular nuclei or share areas of the cerebral cortex such as the somatosensory cortex indicate potential for interaction among modalities.

Visual–vestibular interactions are most notable in the postural stability system because coordinated inputs and feedback among components are essential to good balance. The visual component most involved in balance is the visual periphery where small head rotations (yaw, pitch, and roll) are immediately accompanied by changes in visual flow patterns in the direction opposite to that of head movement.

(The reader can evaluate the role of vision in balance by comparing the length of time balance is maintained when the eyes are opened compared to when they are closed). The visual system throughput appears to be faster in the visual periphery than in the fovea, although the visual system generally is slower than the vestibular system due to the many structures in the central nervous system through which the visual signal must past. The visual periphery appears to play a particularly important role in both postural stability and the perception of self-motion.

Understanding the mechanics of the vestibular system's role in the nervous system is particularly germane. The fact that the vestibular system is connected to neurons controlling eye movements and its position as a fundamental part of the balance system implies that the absence of vestibular inputs might adversely affect the perception of self-motion (Cullen, 2012). Cullen also notes that the somatosensory system as well as the cerebellum are connected to the vestibular nuclei. The result of all of these interconnections is a complex neural subsystem comprising connections from the three main sensory subsystems that serve to support posture stability. Designers need to consider the consequences of removing or degrading information from one or more of these sensory systems.

Control of Posture

The postural control system's neuroanatomy becomes important in understanding how visual imaging systems, whole-body motion platforms, and force cueing devices are likely to combine to influence behavior in vehicle simulators. Massion (1994) in his analysis of the postural control system noted that the postural control system serves two main functions. The first function is to serve as an antigravity control mechanism that supports stance equilibrium. Simply put, the postural control system works against gravity to keep an individual standing upright. When components fail to function, the individual loses balance.

The second function is to serve as a reference frame with respect to perception and action. The postural control system interprets incoming perceptions in reference to the stability of the body. Inputs that signal the possibility of the loss of balance have a high priority and are acted upon without delay. Many of the functions of the postural control system are largely reflexive. That is, they operate rapidly or reflexively at a primitive level. The vestibulospinal reflex (VSR) is one example. The VSR operates autonomously in response to unexpected, positional changes in the body. The VSR is triggered by somatosensory (proprioceptive and tactile) positional cues, which signal undesirable postural changes that occur as a result of the external disturbances in the body. Immediately, return signals are sent from the vestibular system to the musculature to reorient the body. Notably, these reflexive movements of the body signaled from the vestibular system are suppressed during voluntary movements (Cullen, 2012). This might account for the findings that the motions resulting from voluntary vehicle maneuvering control inputs are unaffected by simulator platform motion cues because the resultant inputs from the vestibular system are being actively suppressed by the postural control system.

Disturbance cues are, of course, not limited solely to vestibular inputs. Disturbance to vehicles can also result in proprioceptive and tactile sensations when inertial forces are applied to the vehicle. The forces, however, must be of sufficient intensity to trigger a response. Many low-intensity disturbances are imparted to the body under a variety of conditions. In vehicle control, low-intensity disturbances from environmental factors in both ground vehicles and aircraft are common. These accelerations are brief and of low amplitude. Experienced ground vehicle operators and aircraft pilots routinely ignore these types of disturbances. Even though these disturbances may result in sensations that are suprathreshold, they are not sufficient to trigger a response. Vehicle operators through experience in real-world operations set a criterion for response to disturbance motions that are applied under various conditions. In general, disturbances that do not result in changes to the vehicle state are ignored by the vehicle operator. This is a learned control response to an event and not a function of the postural control system itself. Unlike the case of maneuvering motion, a sensory input from a disturbance is registered but is subject to being consciously ignored.

The postural control system composed of visual, vestibular, and somatosensory subsystems plays a role in how vehicle operators will use available cues in responding to maneuvering and disturbance motions. The vehicle, particularly its dynamics, plays a role in mediating the effect of these forces on the human body. The postural control system will process the resultant sensory input in a way that it would for normal postural control.

Some of the postural control system responses will suppress an active response by the individual; in other cases, a control response will ensue, and yet in other cases higher perceptual control processes will consciously suppress a response.

Control of Locomotion

The control of locomotion, both in speed and direction, is dependent on the same sensory systems that are involved in postural control. Visual, vestibular, and somatosensory systems feedback signals from the individual's interaction with the environment through both the lower-level reflexive controls and higher-level perceptual ones. In an earlier Chapter 3, it was noted how the vestibular system, through its connections with the oculomotor system, automatically moves the eyes in a direction opposite to that of head movement. This synchronization of head and eye movements is purely reflexive. Without actual head movements, there is no vestibular ocular reflex (VOR) as there is no vestibular signal. The eyes can, of course, be moved voluntarily but their movements are not as a result of the VOR. (In fixed-based vehicle simulators without head-slaved visual imaging, the visual motion field is slaved to vehicle movement regardless of the position of the operator's head.)

Movements beyond the range of postural control are involved in normal pedestrian activities. The postural control system still functions under these conditions because balance is still needed. However, the visual, vestibular, and somatosensory systems are now receiving stimulations of different type and magnitude than normally associated with postural control. Initial pedestrian motion results in a flood of signals from all three sensory systems. The vestibular system responds to the initial accelerations of head translational, or linear accelerations if the motion is forward. Rapidly following these signals are sensory transmissions from tactile receptors in the feet and from receptors in the muscles and joints from the first step.

Forward velocity of the body results in optical flow patterns, which the visual system uses to assess both the speed and the direction of movement. Once a constant velocity of movement is achieved, the vestibular signals abate. Only when the velocity is changed while the signaling of the vestibular system will be resumed. Accelerations and decelerations during the movement may also trigger changes in signals emitted from the somatosensory system as the body adjusts to these movement changes. The visual system, however, is not a system that is sensitive to changes in velocity. This seems to be a role specifically suited to the vestibular and somatosensory system. The latter system functions are involved in the voluntary initiation of movement because higher level cortical commands are sent to musculature involved in walking and running. Cessation of that movement will likewise result in similar transmissions. In addition, the somatosensory system is continuously feeding back to higher level cortical areas of the brain the forces imposed by movements of the body on the proprioceptive and tactile systems.

As with the postural control system, multiple sensory data are being transmitted and integrated into a multimodal perception of the body in motion. For pedestrian movement, all the three sensory systems are involved most of the time, though the vestibular inputs would be limited to brief periods of acceleration and deceleration. In the case of vehicle motion, however, some of the sensory system stimuli may be attenuated when compared to the pedestrian vision. Blockage of input from some aspects of the visual scene occurs to varying degrees depending on the vehicle cockpit design. This is particularly true for vertical field of view (FOV) in ground vehicles and many fixed-wing aircraft in which the vertical FOV available to the vehicle operator can be reduced by more than half of what is available to a pedestrian. Vestibular inputs from changes in vehicle velocity are more similar to pedestrians but can easily exceed pedestrian levels. Levels of translation and angular accelerations can be many times higher, especially in high-performance aircraft. In contrast, many somatosensory cues associated with the movement are absent in vehicles because tactile and proprioceptive inputs are now determined mostly by the interaction with vehicle controls, not the external environment.

The designers of vehicle simulators need to be aware of these differences because much of our scientific literature regarding perception is based on pedestrian behavior, a perception unencumbered by the physical constraints of a vehicle cockpit or by the additional mental workload associated with vehicle control. The two major perceptual subsystems described earlier are illustrative of how the human sensory system has evolved to deal with the problems of survival. There are others, but these two are likely to have the greatest impact on vehicle simulator design.

Multimodal Integration of Perceptual Cues

As humans are using multimodal sources of information to maintain postural control as well as to maintain positive vehicle control, it is important to ask how these different sources of information are integrated. From the standpoint of vehicle simulator design, identifying the relative contributions of the different sense modalities is necessary in order to determine design component specifications. If sensory modalities such as vision contribute to substantial amount of information to the support of the task or set of tasks, it would suggest that the design specification be heavily weighted to favor this component. The question then is whether and to what extent these different modalities contribute to the completion of the task. For the designer, this means that some method of quantification of sensory contribution is necessary. One could, of course, build systems that provide the full range of sensory cues and then systematically vary the strength of these cues and assess their contribution to the behavior. To some extent, this is already being done in some research facilities, but the results are limited to a small number of vehicle types and tasks. However, this approach, while having some value in estimating the contribution of sensory cues, is necessarily limited to what these research simulators can provide. Most of the existing research simulators neither can match the foveal capabilities of human vision nor can they match the full range of vestibular responses to sustained linear motion. Furthermore, research simulators are few in number worldwide and cannot be expected to be devoted solely to issues of simulator design. They are often used for other purposes such as determining vehicle handling qualities, designing roadways and airports, and other issues. Finally, research centers may not exist for some vehicles due to the lack of perceived demand or available funding for their construction.

A more viable *analytical* means of determining cue contribution to a task needs to be developed. An analytical method for determining the cue contribution to a task is necessary if perceptual fidelity is to become a practical goal in vehicle simulator design. Specifically, a system of quantification or weighting of perceptual cues both within and across the sensory modalities is needed. This system will inform the design specification for a given vehicle type or class and for each operator task. Thus, equivalency of the weightings for each perceptual cue, for example, between two modalities would imply equal importance in the support of a given behavior. Alternatively, large disparities in sensory weightings favoring as a specific cue would suggest that the design specification should weigh heavily in the direction of the dominant sense. If for example, the production of apparent self-motion is heavily dependent on optic flow with very little dependence on longitudinal inertial cues, the cost–benefit ratio for providing cues to support self-motion perception would strongly favor an advanced display design that can support veridical optical flow patterns. The same design specifications would necessarily deemphasize physical motion cues by providing motion platforms with a relatively small workspace.

Bayesian Perceptual Estimates

The question remains then as to how we can provide accurate estimates of cue weightings both within and across sensory modalities. For this a digression to more theoretical notions of how the human perception functions as needed.

Increasingly, empirical evidence supports the concept that human perception is not deterministic. Rather, the perceptual system operates probabilistically based on what are called conditional probability distributions. In the unimodal case of visual perception, these probabilistic distributions define the likelihood of a visual scene being in a given state (Knill et al., 2003). The distribution is a product, in part, of a noisy visual sensory and perceptual system and in part, the well-structured physical properties of the visual scene. Thus, one can think of the resulting visual image perception as being a collection of these probability distributions with each distribution composed of individual visual cue components. However, the human observer is extracting information from the visual scene to support a given behavior or task. The observer is aided by the fact that the visual world is highly structured even though the observer's sensory and perceptual apparatus are very noisy. As human observers are task-oriented, only a small subset of available information is relevant to support the task at hand. Otherwise, the human visual system would be quickly overwhelmed by a vast amount of information.

The Bayesian model of perception follows an idea of how humans process information similar to that developed for communication theory. The detection of a communication signal is typically accompanied by background noise, so the likelihood of detection is a joint function of the strength of the signal relative to background noise (Green and Swets, 1988). Thus, the likelihood or probability of signal detection is never a simple function of signal strength but always a function of signal strength and background noise.

Bayesian detection theory (BDT), as the perceptual model is called, treats the human perceptual system in a manner similar to the way in which signal detection theory is applied to communications. The perceptual estimate or cue is treated as a probabilistic function of the true state of the world, varying from time-to-time and from individual-to-individual. If the observer operates in a statistically optimal fashion, the most reliable cues or combination of cues are chosen in making the perceptual judgment.

The resulting cues or collection of cues is just one point or perceptual estimate, which has associated with it a distribution of possible estimates. Each such distribution is characterized by a measure of variation (e.g., standard deviation or variance) and a measure of central tendency (e.g., a mean). The larger the variance of the distribution, the less reliable a given perceptual estimate will be for the observer. The Bayesian model can therefore provide a means by which a perceptual estimate or a perceptual cue could be quantified in a way that signal strength can be quantified in communications engineering.

In the Bayesian model, the reliability of the perceptual estimate is determined by calculating the inverse of the variance of the distribution (Ernst, 2006). The reliability of a perceptual estimate r_{rel} can be calculated in the following manner:

$$r_{rel} = \frac{1}{\sigma^2}$$

where σ^2 is the variance of the underlying distribution of perceptual estimates. The inverse of the variance indicates how reliable this estimate would be in supporting a particular task, the higher the variance the lower the reliability. The observer will rely less on a perceptual estimate with low reliability than one with a high reliability.

The measure of human behavior used in constructing these distributions is largely done from psychophysical studies employing methods by which individuals compare the given stimulus (cue) to a comparison stimulus. The resulting distribution of JNDs has a midpoint representing a theoretical point of subjective equality (PSE). That is, a point in which the intensity of any two stimuli is determined by the observer to be the same. Note that because the JND is actually a measure of deviation from a standard, the distribution of JNDs is therefore a distribution of differences. The square of the JND (JND^2) is a measure of variation in the distribution of JNDs.

In the Bayesian model, the JND^2 is a measure of variation in the behavior of the observer. In the Bayesian approach to the weighting of perceptual cues, the inverse of the JND^2 that can be used as a measure of the reliability of the cue, that is, $r_{rel} = 1/\sigma^2 = 1/JND^2$. Using another statistical technique, maximum likelihood estimation (MLE), cues within a single sensory modality or across two or more modalities can be assigned a weight proportional to their reliability when compared to all other available cues. MLE is a common technique when modeling the relative contribution of cues. The weights assigned by the model can be evaluated using statistical regression techniques, which will determine the relative importance of each of a set of cues in predicting performance.

In the case of multimodal cue integration, the Bayesian approach for measuring cues has been evaluated in a number of studies, including Ernst (2006) study of visual and haptic cue weightings and many others. In these studies, systematic variation in the strength of multimodal cues and their effects on behavior was examined. However, for the case of multimodal perceptual fidelity and vehicle control, our concern is only with those studies that examined multimodal cue integration that might directly affect vehicle control.

Multimodal Cues in Self-Motion Perception

The presence of optic flow is a powerful cue to the perception of self-motion and the heading or direction of the observer. In real life, visual motion is usually accompanied by physical motion cues from the vestibular system during observer acceleration and deceleration. By quantifying the contribution of these multimodal cues to the perception of heading, the weighting of each of these cues to the final percept may be obtained. In the study by Butler et al. (2010) predictions regarding the use of perceptual cues were derived using the MLE method. For the multimodal case of combining both visual and vestibular cues, the likelihood distribution is estimated by the following equation:

$$W_{vis} = \frac{r_{vis}}{r_{vis} + r_{vest}}$$

$$W_{vest} = 1 - W_{vis}$$

where:
   W_{vis} is the reliability given by the observer to visual cues
   r_{vis} is the reliability of the visual cue and is equal to $1/JND^2_{vis}$
   r_{vest} is the vestibular cue and is equal to $1/JND^2_{vest}$
   W_{vest} is the weight given by the observer to vestibular cues

The predicted variance of the JND distribution for the combined vision and vestibular cue distributions is given by the following:

$$JND^2_{vis-vest} = \frac{JND^2_{vis} \, JND^2_{vest}}{JND^2_{vis} + JND^2_{vest}}$$

The JNDs were derived from individual observers exposed to visual cues (optic flow) provided by a visual imaging system and vestibular cues (linear accelerations) provided by a motion platform. The behavioral measure was the observer's judgment of the heading (direction) of motion. The longitudinal, physical motion accelerated the observer with a heading congruent to that of the optic flow. If the Bayesian model is correct, then a comparison of heading JNDs should reveal an observer operating in a statistically optimal fashion.

Butler et al. (2010) found in a prestudy that observer response to individual cues produced a JND_{vis} of 6.5° ± 0.6° when only optic flow was present. When the subjects were exposed to only vestibular motion, the JND_{vest} was 7.9° ± 1.1°.

When the subjects were provided with optic flow and vestibular motion, the $JND_{vis+vest}$ for heading was 4.63° ± 0.6°. Note that the heading JND_{vest} is somewhat higher than JND_{vis} heading but the multimodal $JND_{vis+vest}$ heading is substantially lower than with either of the unimodal cues alone. The lower $JND_{vis+vest}$ is the evidence that the combined cues have the effect of improving the reliability of the perceptual estimate of heading beyond that of either cue alone. Calculations by Butler et al. (2010) using the MLE technique indicate that the observer was in fact functioning in a statistically optimal fashion by combining the two cues in such a way as to maximize the observer's judgment of heading.

Although the complete details of the Bayesian approach to the measurement of perception are beyond the scope of this book, the study of Butler et al. (2010) suggest that the methods and measures used in the Bayesian approach may have utility in defining specifications for virtual environments, especially in regard to the contribution of nonvisual design components such as motion platforms.

The previous study by Butler et al. (2010) revealed that the combination of optic flow and physical motion cues significantly enhances observer judgment of heading. It would seem likely that other motion judgment tasks performed by observers would show similar influences of the combined visual and vestibular cues. In a study by Harris et al. (2002), the individual and combined contribution of visual and vestibular cues was assessed for observers who were required to estimate longitudinal distance traveled. Again, the presence of a limited optic flow field (visual cue) alone and the presence of a vestibular cue alone were compared to judgments when both were present. As with the Butler et al. (2010), these observers did not control the apparatus as they would with an actual vehicle but were passive observers who simply provided judgments of the distance traveled.

Harris et al. (2002) calculated the perceptual weights for both unimodal and bimodal (visual plus vestibular) conditions using regression techniques (a form of MLE). Unlike the JND-based method used by Butler et al. (2010), the regression is a means of determining weights of individual cues by curve fitting. The slope of the regression line formed by plotting distance judgments against a target distance for individual cue and cue combination is compared. The regression line or slope is the best fit possible of the judgment data for each cue condition using the linear regression method. The square of this regression coefficient is the coefficient of determination (R^2).

Weights for the various cue conditions were recalculated here as R^2. For the unimodal visual cue only case with a limited optic flow pattern, R^2 was 0.02, whereas the unimodal vestibular cue case resulted in an R^2 of 0.69 for a combined multimodal weighting of $R^2 = 0.71$. The large differences in the weightings of the two studies that are likely due to differences in both apparatus and in the type of judgment being made. Butler et al. (2010) had a visual imaging system with much greater resolution than Harris et al. (2002). As a result the optic flow pattern generated was likely to be much stronger than in the study by Butler et al. (2010). In addition, the tasks required of subjects were quite different, one asking for a heading judgment, the other for a judgment matching a specific distance traveled to a target.

The differences in the results of the studies regarding the relative contribution of unimodal and multimodal cueing in support of specific tasks should be expected regardless of the differences in the apparatus used. Individual unimodal cues are being used by the observer to extract certain types of information from the environment. The weighting assigned to each such cue is proportional to its reliability in supplying that information to the observer. If the task requirements change, it is to be expected that the weightings of perceptual cues will change as well.

FACTORS AFFECTING MULTIMODAL PERCEPTION

If it is true, as seems likely, that observers are optimizing their use of multimodal cues to maximize the benefits of these cues in accomplishing a particular task, then anything that can be done to facilitate the integration of cues would be beneficial. Likewise, avoiding anything that would interfere with their integration should be avoided. A number of factors can affect how an observer's multimodal perception might be affected in a virtual environment.

Cue Congruency

In order to facilitate cue integration and its effective use in multimodal perception, the cues have to be congruent or in agreement with respect to the information they supply. In the Butler et al. (2010) study as described earlier, multimodal integration as measured by the observers' weightings of the value of each cue occurred when the cues provided the same information regarding the direction of self-motion. The cues worked effectively in combination so that the observers' heading judgments with both cues were substantially better than judgments with either cue alone. A second part of this study examined cue integration when the cues provided *incongruent* information, that is, information about heading from optic flow and physical motion that was not in agreement. In this case, the observers were no longer acting in an optimal manner to integrate the cues into a multimodal perception.

Incongruence in perceptual cues can occur for a variety of reasons. System components such as visual imaging and motion platforms may be unable to provide the congruency due to inherent technical limitations. There are often limitations in the precision of the devices involved as well as limitations in the amount of cueing they can provide. System components such as force or motion cueing may simply lack the control accuracy or reliability such that the discrepancies in cueing with other components such as visual imaging results. For example, vehicle operators can extract heading information with an accuracy of about 0.5°; an electromechanical system such as motion platform needs to accurately match this heading at least to the extent that any discrepancies would not be detected by the operator. For these details, the designer should familiarize themselves with the ability of observers to detect differences in perceptual cues such as self-motion heading.

Temporal Synchrony

Temporal synchrony or the synchronization of stimuli in time is important in order to achieve integration of cues. Synchronization in multimodal perception is more difficult and complex than unimodal perception because of the differences in sensory receptor characteristics, transmission times, and areas of the brain involved (Zmigrod and Hommel, 2013). If multimodal cues do not arrive within a specific time period, they may not be perceived as components of the same percept.

In the case of vestibular stimuli, temporal synchrony is especially problematic for virtual environments. This is due to the requirement to physically move the user in a coordinated fashion with visual and other stimuli. Experimental research has indicated that vestibular stimulation has to occur about 160 ms before other stimuli in order to be perceived as simultaneous with them (Barnett-Cowan and Harris, 2009). This appears to be within the temporal synchrony window of 100–200 ms that is required for the support of multimodal perception such as self-motion. Stimuli from different modalities that are outside this temporal window will generally be perceived as entirely separate events.

From the standpoint of vehicle simulator design, providing cues from different modalities within the critical temporal window has been difficult in the past due to problems with transport delays. These are delays due to slow mechanical, electronic, or computational responses such that large amounts of temporal asynchrony were common. A control input could result in a motion platform response delay of as much as 300 ms or more. These large lags between vestibular and other stimuli have serious consequences for user behavior and for the overall utility of the device. These response lags may result in the development of incorrect control skills and even simulator sickness. Modern simulators generally are able to keep these transport delays low and within the temporal synchrony window. Due to the potentially damaging effects of temporal asynchrony, elimination of the offending design component such as motion or force cueing is more desirable than dealing with the problems brought on by temporal asynchrony.

Spatial Synchrony

Multiple sensory inputs can have a spatial as well as a temporal dimension. Spatial congruency occurs, for example, between visual perception of an event and the corresponding auditory stimuli that may accompany the event. In vehicle operations, visually identifiable objects may have accompanying sound stimuli that aids in their localization. For example, certain road surface anomalies, such as rumble strips, are used to warn drivers of road edges. These have clearly defined visual and sound locales, which should be spatially congruent in the simulation. This congruency aids the driver in more rapidly responding to the event. For left-hand drive vehicles, both sound and visual cue will always come from the right of the vehicle corresponding to the right-hand road edge.

Vehicle equipment failures also tend to have multisensory components that are spatially synchronized. These include the sound, vibrations, and the inertial cues associated with tire and suspension failures in automobiles and engine failure in multiengine aircraft. Failure to provide spatial synchrony among stimuli not only reduces perceptual fidelity but can lead to problems in responding to events for trainees. Novice drivers and pilots alike need to learn during simulator training to use multisensory stimuli to localize potentially hazardous events.

Cue Degradation

Temporal asynchrony of cues may result in the perception of two separate events where there should be only one perceptual event. The result is a reduced effectiveness of multimodal stimuli and the potential for introducing false or distracting cues into the simulation. A related problem is the degradation of one or more cues in a multimodal cueing situation. Cue degradation is best illustrated by an example. In aviation where reliance on external visual cues for attitude control is the rule, a pilot may inadvertently fly into meteorological conditions in which attitude information from the visual scene, particularly the visible horizon, becomes degraded. The pilot gradually shifts attention away from the degraded cue to the vestibular stimuli that are normally used to determine the attitude of the body (postural control). Unlike the normally reliable cue to orientation of body position, however, vestibular and related somatosensory force cues can and do lead to false cues regarding the aircraft's attitude. The result is spatial disorientation, a potentially fatal case of unreliable cueing. In real life, pilots are trained to shift immediately to reliance solely on the attitude instruments in the aircraft cockpit and away from cues provided by the postural control system. However, some pilots are either not skilled in instrument flight or transition too late and loss of control results.

In the case of the errant pilot, the reliance on the external visual cues beyond the point where they were providing good information was then compounded by shifting to reliance on a cue, which is inherently unreliable. Other forms of cue degradation occur in ground vehicle operation, particularly in those conditions involving reduced visibility. Forward visibility is especially important for both judging speed and lane position. Degradation of forward visibility means that drivers in real-world operations may have to depend more on the vestibular and somatosensory force cueing for steering control when driving in curves. In addition, reduced visibility may increase reliance on auditory and vibration cueing in controlling speed.

In the case of vehicle simulation, degradation of cues occurs all too often due simply to poor perceptual fidelity. Low resolution and narrow FOV in visual imagery system are common in vehicle simulators with the result that degradation of optic flow fields may force the vehicle operator into excessive dependency on cues from other design components, such as motion or force cueing systems, if they are available.

VEHICLE CONTROL

Multimodal perception plays most important role in the control of the vehicle. Vehicle control can be divided into the same two basic categories that are used for describing the role of a physical motion. Most vehicle control involves maneuvering the vehicle to meet a specific control task such as speed control in ground vehicles or attitude control in aircraft. The other category of vehicle control is a response to unexpected disturbances due to equipment failures or environmental conditions. The former control strategy is an intentional act that uses the available perceptual information in order to carry out a specific control objective. The result of the maneuvering control input is a specific change in the vehicle state, which is continuously evaluated by a vehicle operator. Responding to disturbances however, is generally unplanned and has as its goal to return the vehicle state to what it was before the disturbance occurred.

A further explication of these two types of vehicle control actions reveals how multimodal perception contributes to the vehicle control process. Maneuvering control is an intentional act with an explicit task goal. There is a feed forward aspect to maneuvering control that is absent in a disturbance control act. This aspect of maneuvering control allows the operator to plan for an intended future state to be expressed by a particular control act. The operator briefly stores the state in memory and compares the new vehicle state with it. Continual inputs are made until the vehicle state matches the stored state or criterion.

A change of lane position in an automobile is just such an intentional act in which the vehicle movement from one lane position to another is explicit in the steering wheel input. The steering wheel input has an inner control loop in which the steering wheel forces from the action are fed back to the driver (Figure 5.1). However, recent evidence indicates that inertial (vestibular) cues also play a role during the lane change maneuver (Macuga et al., 2007). These inertial forces resulting from the vehicle directional change can provide the necessary steering guidance even in the absence of visual input from the roadway. Additional somatosensory cues are likely from the shifting of body weight as the maneuver is executed. Thus, the change in vehicle state with respect to the direction is constantly updated by multiple sensory feedback.

Once the amount of control input is decided and acted on, the control feedback shifts to an outer loop consisting of perceptual cues outside the control loop. These include cues from the visual scene, vestibular, and somatosensory forces. In some cases, where control is very aggressive, auditory cues may arise from the steering wheel due to tire friction.

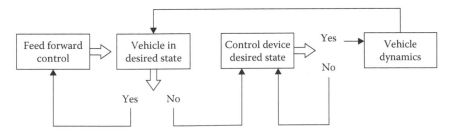

FIGURE 5.1 Feed forward control process for vehicle maneuvering. Desired control device and desired vehicle state require perceptual cues to support control tasks.

The control system of posture has the plant or vehicle dynamics defined by an internalized reference frame, which has evolved to maintain balance. In a vehicle, the dynamics of vehicle control are external to this reference frame, so modulations of control inputs do not come naturally. Instead, the operator learns how to interpret the external cues in reference to the vehicle and, in turn, the vehicle itself controls how the operator inputs will be interpreted and acted on.

Thus, the difference between the system evolved from the control of our postural stability and orientation and that of vehicle control is the additional complication of vehicle dynamics. In this view, vehicle dynamics become, through training and experience, extension of the postural control system. Eventually, the inner loop control becomes largely automated and no longer requires continuous higher level cognitive input from the operator. The outer loop of control takes longer, but it too will become largely automated with experience.

In routine maneuvering vehicle control, the operator is using what might be called a *feed forward* reference frame in which the control of the vehicle is determined by specific intentional task goals such as maintaining lane position and speed in a passenger car or maintaining altitude, heading, and airspeed in an aircraft. In feed forward control the vehicle is essentially the equivalent of the feed forward control for initiating a specific orientation and position of the human body. For voluntary movements, body control and vehicle control will operate similarly with the caveat that the vehicle dynamics can vary widely among vehicle types, whereas for adult human body dynamics it is very similar from one individual to another.

However, there are cases where the normal feed forward control does not apply. When disturbances are applied to the body, such as during a trip and fall, the body's perceptual motor system must react quickly in order to recover stability and position. In this case, the sensory system that functions with the greatest speed is likely to have a large initial impact on regaining stability. In this reactive mode, vestibular and proprioceptive systems are likely to weigh more heavily because of their ability to transmit signals faster when compared to vision and because of their design, which favors detection of acceleration and gravitational forces.

In Figure 5.2, the behavior of a vehicle system responding to a disturbance is depicted. For this example, the task of responding to a wind gust disturbance on a ground vehicle is shown. The vehicle is operating normally with the operator attempting to maintain an optimal lane position. Optimal lane position is the task goal or criterion. Note that the absence of feed forward control as the initiating event here as an external disturbance is, by definition, not anticipated by the driver.

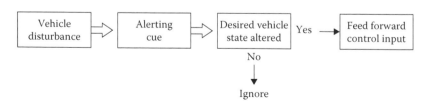

FIGURE 5.2 Disturbance control action to return the vehicle to normal feed forward maneuvering control.

Only after responding to the disturbance does the control of the vehicle return to a feed forward condition. This return to the normal maneuvering control state and its feed forward condition are largely reflexive responses.

At this level of control behavior, the operator is performing a low-level tracking task in response to a temporary disturbance to the vehicle state. The disturbance to the vehicle state has been nulled by a specific control input. Immediately following the disturbance, the operator returns once again to the task of meeting the criterion of optimal lane position.

This journey into the more esoteric areas of vehicle control is intended to reveal the complexities of perceptual processing and to emphasize that the differential weighting of individual percepts is task dependent. The action of the feed forward-based maneuvering control action is in contrast with the reaction of vehicle control that results from disturbances. Within each of these types of control actions, certain perceptual systems are likely to dominate others because they are providing more reliable and relevant vehicle control information to the vehicle operator.

In the case of maneuvering control, operators are actively seeking information that supports the task. Seeking and extracting information from the external environment means that sensory resources that are likely to be used in this action are primed for the reception of information. Visually, the operator is attending to cues that are already known to be supportive of tasks such as speed control and lane position. Thus, the threshold for these percepts is likely to be optimized. The feed forward mechanism primes the relevant sensory systems for reception. The visual system orients the eyes and adjusts the focus in order to prepare for the reception of data from the road ahead.

Contrast this voluntary maneuvering activity with the responses to vehicle disturbances. In the absence of alerting information (warnings of equipment or environmental disturbances to come), the operator is wholly unprepared for the event. The usual anticipatory responses are absent and the operator is attending to an information stream that is unrelated to the impending disturbance. It is therefore likely that the perceptual event that accompanies the disturbance cue may be described as having sensory components that are not primed for the event. The thresholds for perceptual response to this event may be higher than to an anticipated event because attention is being allocated to sensory systems unrelated to the event. For example, the linear accelerations accompanying a gust disturbance event on a roadway and the same type of cues from an aircraft engine failure on takeoff are events that will likely require higher accelerations to exceed operator threshold than the normal control responses in which the operator is anticipating vestibular feedback.

Implications for Design

Humans are confronted with an overwhelming array of sensory information and daily activities. From this milieu, information in the form of perceptual cues is selected to support the task at hand. Most sensory information received by our visual and other systems does not reach higher levels of processing. If it did, we as humans would become paralyzed by excessive information. Instead, we select the information needed by orienting our senses in two fundamental ways to increase the likelihood of information reception. First, we can physically orient our senses. In vision, we do this by orienting our head and eyes to the location of the desired information. We also focus our attention or prime our perceptual system in anticipation of the incoming information. We do this routinely in an aural communication in which we block out the relevant sounds and focus on the communication at hand. The heightened attention to the task enhances the performance by priming a system to action.

Humans are task-oriented in their behavior. Even if they appear to be inattentive to the task at hand, it is because they are attending to another task that may be purely cognitive in nature. For the design of any virtual environment, a detailed understanding of the operator's tasks, especially the perceptual and perceptual-motor tasks, is essential for cost-effective design.

The requirements for a cost-effective design can only be achieved if the perceptual cues necessary for the successful completion of a task are provided to the user. This means not only that these cues are available, but that they are represented in a virtual environment in such a way as to maximize their effectiveness. It is evident from the perceptual literature that individuals weigh the value of the information that they receive from the environment in proportion to its inherent reliability in supporting perceptual judgments.

SUMMARY

Multimodal perception presents a particular challenge for the design of virtual environments due to the difficulty of isolating the relative contributions of individual cues that are used in each task. In this chapter, the evidence that individuals weigh the contributions of each cue in relation to their perceived reliability in supporting perceptual judgments was provided. In addition, quantitative methods were described that allow designers to measure the relative weights for these cues in order to determine their value in the design. Attributes of multimodal perception that affects the validity of cues were described, including cue congruency, temporal synchrony, spatial synchrony, and cue degradation. The potential impact of poor perceptual fidelity with specific regard to cue degradation as well its impact on behavior was described.

REFERENCES

Barnett-Cowan, M. and Harris, L.R. 2009. Perceived timing of vestibular stimulation relative totouch, light and sound. *Experimental Brain Research*, 198, 221–231.

Butler, J.S., Smith, S.T., Campos, J.L., and Bulthoff, H.H. 2010. Bayesian integration of visual and vestibular signals for heading. *Journal of Vision*, 10, 1–13.

Cullen, K.E. 2012. The vestibular system: Multimodal integration and encoding of self-motion for motor control. *Trends in Neuroscience*, 35, 185–196.

Ernst, M.O. 2006. A Bayesian view on multimodal cue integration. In G. Knoblich, J.M. Thornton, M. Grosjean, and M. Shiffner (Eds.), *Human Body Perception from the Inside Out* (pp. 105–131). New York: Oxford University Press.

Green, D.M. and Swets, J.A. 1988. *Signal Detection Theory and Psychophysics* (rev. ed.), Los Altos, CA: Peninsula Publishing.

Harris, L.R., Jenkin, M.R., Zikovitz, D., Redlicak, F., Jaekl, P., Jasiobedzka, U.T., Jenkin, H.L., and Allison, R.S. 2002. Simulating self-motion I: Cues for the perception of motion. *Virtual Reality*, 6, 75–85.

Knill, D.C., Kersten, D., and Yuille, A. 2003. Bayesian models of object. *Cognitive Opinion in Neurobiology*, 13, 150–158.

Macuga, K.L., Beall, A.C., Kelly, J.W., Smith, R.S., and Loomis, J.W. 2007. Changing lanes: Inertial cues and explicit path information facilitates steering performance when feedback is removed. *Experimental Brain Research*, 178, 141–150.

Massion, J. 1994. Postural control system. *Current Opinion in Neurobiology*, 4, 877–887.

Zmigrod, S. and Hommel, B. 2013. Feature integration across multimodal perception and action: A review. *Multisensory Research*, 26, 143–157.

6 Quantifying Perceptual Fidelity

INTRODUCTION

Designs of virtual environments that are driven by the requirements of physical fidelity only require that the design will in some way match physical reality. The visual imagery system must produce photorealistic scenes at a level that matches or physically replicates the real-world image. Similarly, the physical forces imposed on a passenger vehicle maneuvering around a curve of a given radius at a given speed can readily be calculated. These forces, in turn, are then replicated in simulation. There is no need or requirement to specify anything about how the driver of the car will perceive these motions. The designer simply has to specify what type of component (e.g., motion platform) performance is needed to provide these accelerations and for what duration.

The designer pursuing the goal of perceptual fidelity, on the other hand, must choose from several possible options or methods of how to calculate the perceptual fidelity for a given design component. The choice of each of these carries with it some degree of risk as well as cost. The risk may be the development of specifications which are far in excess of what is needed for the device. With this overengineering comes both high cost and potential failure in the marketplace. Alternatively, the specification may fall short of what is needed and though attractively priced, it fails to deliver value to the customer. In this case, the device is not capable of supporting the critical user behavior needed to perform the desired tasks.

SENSORY FIDELITY

A virtual environment designer that wishes to be ensured that the design component will provide at least a minimum amount of perceivable stimulus to the user can opt for a design strategy that is based on suprathreshold stimulation. This is a *sense-limited* stratagem, which might be more accurately described as sensory fidelity because it is simply defining a minimal stimulus level that will trigger a user's sensory receptors. Sensory fidelity is not task-dependent and does not require a specification of the types of task that the user of the device might perform. Rather, the specifications are based purely on the ability of the system components to meet or exceed specific human sensory receptor stimulus values.

The most common example of this type of sensory fidelity is the use of the term *eye limited* when referring to the level of detail in an image. Eye limited means that the perception of a visual display detail is limited by the visual acuity of the user.

This typically is meant to imply a Snellen acuity of 20/20 for the user. That is, a display with 1 arc min/pixel measured at the design eye point. Note that the intent of this specification is to match the level of display resolution to that of the maximum visual acuity the user might possess.

Similarly, image contrast levels to meet these sensory fidelity criteria would need to be at least 8:1 (Wang and Chen, 2000) as this has been found to be the minimal contrast level needed to optimize user performance. Again, these values would meet or exceed the contrast sensitivity of the human eye with regard to image displays. Once again, this is not about the particular task involved but rather a general specification as to the minimum requirement of image contrast that is needed to carry out any task. There are cases where objects may exceed these levels under certain conditions such as headlight glare. However, sensory fidelity does not address specifics regarding certain conditions or tasks but only sets design specifications with regard to the sensory receptor capability of the intended user.

Image Movement

Image frame rates were originally established by the film industry in order to assure that the moviegoers would see continuous smooth motion. They established a standard frame rate at 24 fps. This is, in fact, a very early instance of sensory fidelity because it exceeds the critical threshold flicker frequency of humans. Critical flicker frequency is the threshold at which a user begins to a see a flickering image rather than a continuous image with smooth object motion. The resultant flicker was found to be unacceptable by audiences, so a standard frame rate was set to exceed this frequency and present the illusion of continuous motion. Most virtual environment imagery systems use 30 fps as the basic rate but many exceed that number particularly if high speed motions are to be displayed.

Field of View

The instantaneous horizontal field of view (FOV) available to adult users is about 180° with a vertical FOV of about 110°. Providing display imagery at this level takes advantage of the full range of retinal receptors that are available to both eyes. The designer would want to provide at least 300 cd/m^2 of luminance to assure that the retinal receptors were functioning at an optimal level.

Color

The color display specifications for sensory fidelity are somewhat more difficult to calculate due to the wide variety of estimates in the literature. This is due to the difficulty of conducting pairwise comparisons over such a large number of colors. If we use the most reasonable estimate for color discrimination then only about 7000 would be needed, far less than the more than 16 million available for the typical 24-bit true color display system.

Vestibular

Motion platform design specifications, particularly those involving Stewart platforms, typically set the threshold levels for angular and linear accelerations that correspond to the motion detection thresholds in the published literature. This is of great importance when the actuator stroke needs to be returned to its neutral position. If the actuator stroke return is too fast, it may exceed the motion detection threshold of the operator of the device resulting in false and possibly disorienting motion cueing. This is yet another case in which sensory fidelity has been employed as a design goal at least with respect to this particular design component.

Published levels of angular acceleration thresholds will vary depending on the method used. The lowest recorded angular acceleration thresholds averaged to $0.11°/s^2$ (Clark and Stewart, 1968) and the lowest average threshold for linear acceleration recorded was 0.01 m/s^2 (Gundry, 1978). Note that these thresholds are measured under ideal laboratory conditions designed to determine the minimal level of vestibular stimulation required to elicit a response.

Somatosensory

Applying the same sensory fidelity design strategy to proprioceptive and tactile sensory receptors requires a more nuanced approach as different areas of the body prospect on two different levels of stimuli. The distribution of tactile mechanoreceptors varies dramatically across the skin surface of the body. Moreover, in both areas, data on basic discriminative ability is relatively limited when compared to other sensory systems such as vision.

For the cutaneous mechanoreceptors at the surface of the skin, there are two measures relevant to vehicle simulator design: (1) pressure and (2) vibration. Passive pressure thresholds on the hand show an average threshold of 0.04 g for young adults increasing to 0.16 g for the elderly (Bowden and McNulty, 2012). Tests of passive sensitivity to vibrations on the hand show a wide response variability particularly at low levels of vibration (<25 Hz, 1 mm peak to peak amplitude) due to differences in skin thickness. The differences are generally eliminated at higher levels (>100 Hz) (Lovenberg and Johansson, 1984).

Differentiating between passive and active sensory function is important in quantifying the likely response profile of virtual environment users. In the case of passive sensory threshold, the individual is likely to be attending to other incoming information or other tasks and this fact may mean that higher levels of stimulus energy are needed to exceed threshold will be needed than in the active processing case. In the case of active sensory function, the individual is attentive to the specific sensory input and this will most likely reduce the stimulus imagery required to elicit a response. For setting threshold values that assure a response by the observer, the choice should be a higher value than that associated with passive attentiveness.

Perceptual Discrimination and Perceptual Fidelity

Quantifying sensory fidelity on the basis of the limitations or response profiles of human sensory receptors is relatively straightforward. One needs only to determine the parameters of the particular sensory receptor and set the design specification to exceed that level. In some cases, these parameters are based on the physical attributes of the receptor such as visual acuity. Other parameters are based on threshold responses to various levels of energy. Admittedly, there will be virtual environments such as vehicle simulators, which may have tasks in which values of stimuli need not be as high as that determined by sensory fidelity. However, if the virtual environment under development is intended for very wide applicability, then setting the design goal at this level might be reasonable. For example, a virtual reality system consisting of a HMD visual imagery system and an associated auditory sub-system may be developed as a device with potentially universal applications. In this case, setting the visual display resolution as eye-limited, for example, 1 arc min/pixel is justifiable as there are no specific task requirements which could guide a more precise set of specifications.

To establish a more accurate level of perceptual fidelity, however, a more refined view of human response to external stimuli is needed. Psychophysics initially developed methods in which simple thresholds to stimuli could be determined. Psychophysical methodology eventually went beyond the establishment of a simple threshold response to the presence of stimulus energy to measure the observer's ability to discriminate *differences* in energy levels of a stimulus. The result was a set of quantifiable, psychophysical relationships, which is more descriptive of perceptual processes than that of simple sensory receptor thresholds.

The Just-Noticeable-Difference

Typically, psychophysical methods will rely on a force choice methodology. For a threshold responses, the subject simply states whether a stimulus is present or not a stimulus or there was an absence of a stimulus. Sometimes the stimulus will actually be present, sometimes it will not. Adjustments are made to the level of stimulus energy so that a value can be determined at which a stimulus is judged to be present half the time. This is the threshold value. In the case where a comparison is to be made between two stimuli, a choice is required as to whether the intensity of one stimulus is greater or less than another. A standard stimulus is presented followed by a comparison stimulus, which varies in value in a predetermined manner. The observer is required to judge whether the comparison stimulus is greater or lesser in value to the standard. This psychophysical method generates a distribution of just noticeable differences (JNDs) that have been discussed in earlier chapters.

In the case of the threshold, the stimulus energy necessary to exceed the threshold is stated. It may be a value of light energy emitted or sound energy level, which the observer detects at a level above chance. Note that the threshold is an average either of an individual subject or a group of subjects. Some subjects will have a threshold somewhat lower or somewhat higher than this average level of energy. If the designer wants to make sure that most of the user population will sense the stimuli than a stimulus energy level greater than that threshold should be used. This suprathreshold would be set at a level that is likely to be detected by the vast majority of users, perhaps 95% in the more extreme case. Assuming an underlying Gaussian or normal distribution of thresholds for a particular stimulus, then a stimulus value of two standard deviations above the threshold will be detected by 95% of the population. If the stimulus thresholds distribution is not Gaussian, knowledge of the actual distribution of values will be needed to set the value.

In addition, even a more useful measure than the threshold is the JND. More often than not, the designer is likely to be interested in how well the virtual environment user can discriminate between one value of a suprathreshold stimulus and another.

The JND can also be used to determine whether the user will be able to discriminate between the stimuli presented in the virtual environment and a similar stimulus presented in the real environment. For example, in making subjective judgments regarding the realism of a virtual environment, the user is always relying on memories of real-world stimuli. If the differences between the stimuli presented in the virtual environment are within the JND of the real-world stimulus, it is unlikely that the differences between the two environments will be detected.

In some cases the JND for a given class of stimuli is so consistent across the range of stimulus values that it takes on a lawful relationship. Weber's law or Weber's fraction states that the detectable difference (JND) between the two values of a stimulus is always proportional to the magnitude of the initial stimulus or stimulus used as a standard. If the magnitude of the stimulus is small, the difference needed to detect a difference in stimulus energy will be small as well. Likewise, a large difference between stimuli will be needed if the initial stimulus magnitude is large. This law applies generally to all types of stimuli, including light, sound, pressure, physical motion, and others.

The JND has a wide application within both neurological and psychological disciplines. A JND can be calculated for light or sound energy or the pressure or vibration difference detection on the skin. It has been also used to determine the relative value of cues to distance such as relative size or relative (texture) density (Cutting and Vishton, 1995).

There are other means of measuring the perceptual performance of individuals whether in virtual environments or in the real world. Application of signal detection theory or SDT (Green and Swets, 1988) has been used as means of assessing perceptual performance as it allows the calculation of signal detection judgments independent of an observer's criterion for making that judgment. Inevitably, any judgment includes a criterion for making a perceptual judgment. Some observers may be more conservative in setting the criterion for judgments and are unwilling to make these judgments unless confidence in the judgment is high, whereas others are much less conservative and are willing to make more errors in judgments. However, very few studies in the perception literature use the SDT methodology making this a less likely candidate for perceptual fidelity measures.

The use of psychophysical methods is one of several approaches to quantifying perceptual fidelity as well as the perceptual capabilities of users. As with sense-limited measures that are used to achieve sensory fidelity, the use of threshold and JNDs derived from psychophysical methods is useful for describing fundamental characteristics of human perception. Many of these characteristic perceptual limits and capacities support more complex behaviors found in the operation of vehicles. As will be seen in Chapter 7, the designer needs to understand what those perceptual elements are that support each of the tasks that the vehicle operator is to perform in the real world. Then a determination needs to be made whether the particular simulator design component, such as the visual imaging system, will provide the necessary perceptual cues to the support of each task. Using perceptual thresholds, the designer can determine the minimal value a perceptual cue needs in order to be detected by the operator. The use of the JND or a similar measure of user perceptual ability can then be used to determine whether design specifications are within the range in which the operator cannot detect a difference between the stimulus event in the real world and that of the simulator. Regardless of which measures are used, the designer needs to be aware that these measures are just estimates of the average perceptual performance of observers. Individual performance will vary around these average values and that variation should be considered when evaluating the performance of simulator users.

Bayesian Perceptual Estimates

The use of basic measures of the perceptual performance of observers, the JND, has also been used in modern perceptual theory regarding how observers weigh the reliability of the perceptual cues. Probability theory, particularly probabilistic functionalism, has played a role in perception theory for decades (Brunswik, 1955 and others). This is the idea that observers optimize their perception of the world by weighing the utility of the perceptual cues based on the predictive value in some ecologically valid outcome. The stronger the predictive value of a cue (e.g., optic flow) to an important perceptual judgment (e.g., self-motion) needed for survival, the higher is its ecological validity.

The idea of perceptual estimates using the Bayesian theorem was discussed in Chapter 5 in relation to the weighting of cues in multimodal perception. The use of the Bayesian model can be extended to include a broader spectrum of cues and even as a means of estimating perceptual fidelity of a virtual environment. In Bayesian perceptual theory, each percept or perceptual event or estimate is a sample from a distribution of preexisting estimates (or priors) experienced in the past combined with an estimate based on the current (posterior) perceptual event.

The perception of an observer for a given event is a perceptual estimate that is based on the combination or product of individual perceptual estimates that contribute to the perceptual judgment. Experienced vehicle operators have a rich store of perceptual experiences to draw on some of which appear to be prewired at birth, as with some depth cues, whereas some are learned over years of interacting with the environment. In the case of novice drivers, the basic cues of interacting with the environment such as the perception of speed distance already exist. The novice driver now needs to combine the cues stored in memory with the incoming information regarding speed and distance. Due to the high speeds experienced in driving, modifications of the existing representations and relationships regarding speed and distance will be needed. It is these modified representations that will be the best predictors of behavior in the future. Not until the novice driver has achieved driving proficiency, however, can their experiences in driving simulators be used to validate the perceptual fidelity of the device or any component of the device because their perceptual estimates do not yet have predictive validity. The same applies for the perceptual-motor tasks that the novice needs to learn in driving a vehicle. The novice will know the relationship between the body motor control and the changes in perception of the environment, but not the relationship between a vehicle control input and the same perception. These perceptual-motor representations are learned with repeated exposure to these vehicle inputs and their consequences.

As noted in Chapter 5, the Bayesian perceptual estimates are based on the variability of the distributions that underlie the estimate. As a given perceptual estimate is just one point on the distribution of estimates, the estimate has variability associated with it. As a sample from any distribution, repeated sampling will result in a distribution of slightly different samples. Thus, each distribution of perceptual estimates has its mean and a measure of variability. A perceptual estimate or cue to the distance of an object from the observer is just one sample from a distribution of estimates. The reliability of the estimate increases if the distribution has a low variance, decreases if the variability of the distribution is high. This inverse relationship between the reliability of perceptual estimate or cue was expressed earlier by the expression $r_{rel} = 1/\sigma^2$. To convert the reliability measure to a useful quantity requires an estimate of the underlying distribution of the cue. A viable candidate for this is the JND or, more accurately, the JND^2.

Another component of the Bayesian model is that the observer operates in a statistically optimal fashion when there are multiple cues available. The observer will combine these cues in a manner that maximizes the accuracy of the resulting perceptual judgment. This is done by converting the reliability of a cue to a weighting of that cue in making the perceptual judgment. The higher the reliability of the cue the more likely the observer will rely on that cue to make the perceptual judgment.

For a combined estimate of a perceptual judgment, the reliability of the combined cues (r_{tot}) is found by summing the reliabilities of the separate cues:

$$r_{tot} = \sum r_i$$

In the situation in which there are multiple cues, as is common in the real world, a reliability estimate is needed for each cue involved in the perceptual judgment. As the reliability of a cue is inversely related to its variance, each reliability of the cue can be calculated by using the inverse of the estimate of its variance, that is, $1/JND^2$.

Substituting $1/JND^2$ with r_i to simplify the expression, we can calculate the reliability that an observer assigns to any cue, provided we have an estimate of the variability of the distribution from which it came, such as the JND^2.

Applications to Design

One might ask at this point what role the Bayesian model of perception has in designing a virtual environment and how it might contribute to the quantifying of perceptual fidelity. The details of how this might be accomplished in a practical setting will be discussed in Chapter 7. However, it should become immediately apparent that achieving a high degree of perceptual fidelity requires an understanding of how various perceptual cues from various sensory modalities contribute to a given perceptual judgment such as distance, speed, and spatial orientation. For any given task conducted in a virtual environment, the calculation of the reliability estimates of these cues is a means of determining their value to the observer.

The use of these Bayesian estimates or weights in design becomes clear when we examine the role of perceptual cues as a collective system for supporting a task, whether the task is performed within a virtual environment or in the real world. In the Bayesian framework, the observer is operating in a probabilistic fashion in the use of perceptual cues. The goal of the observer is to integrate the various cues in a statistically optimal fashion based on the reliability assigned to each cue.

Note that we are assuming that the individual has constructed an integrated cue collective for this particular task. The task may be supported by one or more perceptual cues within a single modality (such as vision) or one or more such cues from each of the two or more modalities (e.g., vision plus vestibular). These reliability estimates assigned to individual cues can be used by the designer to prioritize the values of each cue for the virtual environment user for any given task. Thus, the design specifications are value-driven in that they are based on their weight in supporting a task.

As an illustration of how perceptual judgments and therefore perceptual fidelity can be quantified, perceptual weights were calculated from the Cutting and Vishton (1995) data discussed in an Chapter 2. One cue, object occlusion (also known as object overlap) was excluded from the dataset due to its limitation to the relative distance of object pairs rather than the more useful egocentric distance judgments of individual objects. A second cue for distance, linear perspective was not analyzed by the authors. They considered linear perspective as a system of cues and not a cue itself. However, components of the linear perspective system such as relative density are included in the analysis.

Note that distance cues are not constant in their effects on distance judgments and, in fact, in the Cutting and Vishton (1995) analysis only relative density and relative size provide consistent cue value regardless of the distance of the viewer from the object.

The weights are derived based on the proportional reliability of the cue. Thus, although individual variance estimates of the cue are derived from the inverse of the JND^2 ($1/JND^2$), the weight itself is determined by calculating its relative contribution to the estimate when compared to the other cues available at that distance. The weight is expressed as the following:

$$w_i = \frac{1/JND^2}{\sum(1/JND^2)}$$

where:
   w_i is the weight assigned to a given cue for a specific task
   $1/JND^2$ is the reliability estimate of that cue

The sum of the combined weights for all of the perceptual cues supporting that perceptual judgment must equal 1 or unity.

In Table 6.1, weights for the individual distance cue JNDs provided by Cutting and Vishton (1995) are given as a function of the distance of an object from the observer. Note that some cues are not available at all distances. The original dataset included ranges from 0.5 m to 5 km. However, aerial perspective and relative size cues dominate the distance perception beyond 1000 m, whereas distances below 2 m were not deemed relevant to the issue of distance perception as it applies to the generation of visual scenes in vehicle simulators.

One of the distance cues in the Cutting and Vishton (1995), relative size has a constant JND of 0.03 (3%) regardless of the distance. In the absence of any other cues,

TABLE 6.1

Perceptual Weightings of Distance Cues as a Function of Distance from the Observer

Distance Cue	Egocentric Distance					
	2 m	**5 m**	**10 m**	**30 m**	**100 m**	**1000 m**
Binocular disparity	0.316	0.019	0.033	0.027	0.002	–
Height in the visual field	–	0.633	0.817	0.436	0.157	0.001
Motion perspective	0.645	0.310	0.051	0.061	0.039	–
Relative size	0.035	0.034	0.091	0.436	0.736	0.766
Relative density	0.003	0.003	0.008	0.039	0.066	0.069
Aerial perspective	–	–	–	–	–	0.163

the relative size cue is highly reliable. However, note the change in weighting of the relative size cue when other distance cues are included.

The Cutting and Vishton (1995) analysis of distance judgment cues was an assessment of how visual cues are used to judge the relative distance of objects in a visual scene. By using the Bayesian model, reliability estimates and relative weights are calculated from these data to determine the relative utility of each cue as a function of egocentric distance. This approach provides a more accurate assessment of the value of cues as it allows a comparison of each cue in relation to others that may be available at that particular egocentric distance. For example, the weighting of the binocular disparity cue drops dramatically from 0.32 at 2 m to 0.019 at 10 m. The precipitous decline of the binocular disparity cue reflects its low reliability at the greater distances when compared to other distance cues that might be available.

It should be evident at this point that quantification of perceptual fidelity requires a more detailed level of analysis with respect to the information available in the task environment than the virtual environment designer has come to expect. This requirement for in-depth task decomposition, to be discussed in more detail in Chapter 7, becomes even more evident when calculating the influence of cues from different modalities on perceptual behavior.

Quantifying Multimodal Cues

The Cutting and Vishton (1995) data describes the relative contribution of cues within a single modality: Vision. Many tasks, however, involve the use of two or more sensory modalities. As discussed in Chapter 5, multimodal cues can be quite complex and can combine in different ways for different tasks. Understanding how different modality cues influence operator behavior allows the designer to estimate the relative importance of each to the simulation design and the cost-effectiveness of their inclusion.

The measure of the reliability of each cue for each sensory modality can be calculated. In the case of multimodal cueing, the formula for estimating the weight assigned by the subject to the visual cue is as follows for the case of the combined visual and vestibular cue:

$$W_{vis} = \frac{1/JND^2_{vis}}{1/JND^2_{vis} + 1/JND^2_{vest}}$$

where JND^2_{vis} and JND^2_{vest} correspond to the variance of the visual cue and the variance of the vestibular cue, respectively. Once again, the reliability of each cue is the inverse of the variance of that cue and the weight of the cue is the reliability of the cue in proportion to the total reliability of the cues available to the user in the conduct of that task.

Effect Size and Cue Weighting

Thus far, only the use of the JND has been discussed with respect to the weighting of perceptual cues in the design of virtual environments. A JND is a measure of the degree perceptual discrimination of a particular stimulus. It is used here as a measure of the reliability of a cue when used as a measure of variance ($1/JND^2$). In the same way that the cue weight is a measure of the relative importance of a cue, the effect size in an experimental study of perception in vehicles is a measure of the variance of behavior due to the variance of a cue or a combination of cues. However, the effect size in these studies is a direct measure of the effect of the perceptual cue, such as optic flow, on vehicle control behavior. Many of these studies have already been cited in this book. When available, effect sizes can be used as a means of assessing the relative importance of a cue. For this reason, it is important to understand how it is derived and how it can be used as a measure of perceptual fidelity.

An experimental evaluation of the efficacy of perceptual cueing in vehicle control measures the effect on control behavior, such as speed control, which results from the changes in the intensity of some perceptual cue, such as optic flow. In the experiment, the average vehicle speed of a group of drivers tested with a visual image having high optic flow would be compared to a group of drivers tested with a visual image having low optic flow. The difference in group means is evaluated with a test of statistical significance to determine whether the difference between the means is due to optic flow or just to random sampling error. The term statistical *significance* may be misinterpreted as *important* where, in fact, the importance of the finding has nothing to do with the statistical tests for sampling error. A finding of statistical significance in an experimental study is about the reliability of the finding not about its importance. Moreover, the finding of statistical significance has nothing to do with the *size* of the effect either. To determine the size of the effect, it is necessary to do a separate analysis.

The measure of the size of an effect, the effect of perceptual cue on the driver behavior in this example, is to compare the variance of the driver behavior due to the cue against the total variance of the driver behavior measured in this study. The result is a measure of the strength or the effect size of the perceptual cue expressed as a single number such as a proportion or a percentage.

There are several measures of effect sizes available. One of which, R^2, has been mentioned in Chapter 5 as a general index of the effect of a perceptual cue on behavior.

It is often used in correlational studies or in studies attempting to fit data to a particular model. Another effect size measure often used in experimental studies, eta-squared (η^2), is calculated from the analysis of variance statistical test (also called the F test). It is the ratio of the mean square treatment to the mean square total: MS_{treat}/MS_{total}. If there is more than one treatment variable in the study, separate values can be calculated for each of these or any combination of treatments. (One caveat is that designs using subjects repeatedly in treatment conditions will tend to inflate the effect size unless a correction is used).

Another measure of effect size is commonly used in meta-analytic studies. These studies combine or pool the variances across several studies and sample sizes in order to measure the pooled or cumulative effect size of variables. The measure of the effect size, Cohen's d, has the advantage of a pooled sample size, which results in more power in testing the results for statistical significance. In vehicle simulation cueing effectiveness, Cohen's d has been used most recently in the meta-analysis by deWinter et al. (2012) on the effects of motion cueing in flight simulators. The deWinter study examined motion platform cueing affects across a variety of aircraft simulator studies and motion cueing conditions. Prior to this study, no individual study of motion cueing measured by training transfer showed any statistically reliable findings. Pooling the sample size and variances of the studies allowed deWinter to assess the role of platform motion cueing in a manner that is more favorable to statistical testing. The meta-analyses revealed statistically significant differences in which none had been found before.

Designers, who encounter the Cohen's d, while useful for assessing treatment effects, should be aware that the measure can be somewhat misleading. This is due to the fact that the pooling of variances across several studies means that a single study with very large differences in treatment effects may distort the findings. In the case of the deWinter et al. (2012) study, motion cueing was found to be effective for disturbance effects for combined fixed wing and helicopter. It is possible, however, that the effects are due to differences in helicopter control behavior due to motion cueing effects, rather than any effects for fixed-wing aircraft. Therefore, caution should be exercised in the interpretation of the findings of meta-analytic studies.

These three measures of effect size are the most commonly used in vehicle simulation studies, but there are others. Each such measure can be converted into the other mathematically. Each has its advantages and disadvantages but all ultimately yield the same results. Unfortunately, many studies in applied settings, including many studies of vehicle simulation design effectiveness, do not report effect size or even the variance estimates that could be used to derive the effect size. Thus, the designer is left unsure whether a statistically reliable effect is or is not an important effect when deciding on what design trade-offs need to be made. A statistically reliable effect of a perceptual cue that has a small effect of only 6% means that 94% of the variance in the behavioral measure is due to something other than the perceptual cue. For example, if the effect of motion parallax produced by a head-slaved visual imagery system is found to be statistically reliable, the implication is that it might be a requirement for vehicle simulators if the effect size were relatively large, for example, 67%. However, the design judgment might be quite different if the effect size was determined to be only 6% of the total variance of the study.

Determining the effect size for an individual study is only possible if the treatment effect, if any, has been determined to be statistically reliable. Only when the treatment effect size exceeds a specific level relative to the total variance in the study is it meaningful to discuss the issue of effect size. If the effect size does not exceed a specific level, it is likely that these effects are due to random sampling error. In other words, no *effect* really exists.

The advantage of calculating effect size in studies evaluating the value of a particular simulator design component is that it provides a direct measure of not only whether the component makes a difference in behavior, but the size of that difference as well. Ideally, the study has evaluated differences on behavior that has high external validity such as the control of vehicle speed or similar measures. Effect size calculations are still relatively rare in real-world field studies of perception due to the difficulty of systematically varying perceptual cues in real-world conditions. Most of the effect sizes of perceptual cues that do exist were derived from studies conducted in research simulators and not real life. When derived from simulator studies, designers should carefully examine the capabilities of the simulators that are used in the study in order to determine the degree to which confidence in their results is warranted.

Some examples of effect size data derived from vehicle simulator studies are shown in Table 6.2. These data are from both ground vehicles and aircraft studies in which design components affecting the vehicle operator perceptual behavior were found to be reliable and in which an effect size was calculated by the authors or could be derived from the data reported. For the sake of clarity, the original effect size calculations are converted to R^2. The use of the squared correlation coefficient is generally easier to understand as the correlation coefficient is simply a measure of the degree to which variation in a perceptual cue such as optic flow is associated with variations

TABLE 6.2
Effect Sizes of Perceptual Cues on Vehicle Operator for Ground Vehicles and Aircraft. Effect Sizes as Stated are R^2

Source	Vehicle	Task	Cue	Behavior	Effect Size
Lee and Bussolari (1989)	Large transport	Approach landing	Motion pitch	Pitch control	0.26
Ricard et al. (1981)	Helicopter	Hover	Motion pitch/roll	RMS deviation hover point	0.05
Beukers et al. (2009)	Small transport	Decrab landing	Motion roll	Roll rate control	0.49
Zaal et al. (2015)	Large transport	Sidestep landing	Motion vertical	Sink rate	0.53
Palmqvist (2013)	Passenger car	Passing maneuver	Head-slaved motion parallax	Lane position	0.06
Diels and Parkes (2010)	Passenger car	Speed production	Optic flow	Speed	0.31

in vehicle behavior such as speed. It varies, unlike the simple correlation coefficient, from 0 to 1.00. There are no negative effect sizes using R^2.

The first study cited by Lee and Bussolari (1989) showed a significant negative effect of motion cueing on pilots with no experience in flying large transport aircraft. Effect size measures do not show the direction of the effect of a cue, only the size of the effect. The Diels and Parkes (2010) effect size for speed production reveal the strength of optic flow effects on the perception of self-motion.

Primary and Secondary Cues

It is often the case that some cues and some modalities, in general, seem to dominate the weighting of cues. For applications such as the design of driving simulators, the predominance of visual cues is not surprising as driving is primarily a visual task. The tendency is to ascribe dominance to certain perceptual information without the evidence that a cue or modality predominance should take priority in the design specification.

Quantification is achieved when the specific cue or set of cues supporting a particular task is specified and a weight or measure of its utility is provided. That utility is defined by the cue's effect on task behavior relative to other cues that may be available to support that particular behavior. Only at this point, is there sufficient information to support a design specification.

In order to assure that the most important cues are given primacy in the design specification, a system of rank ordering of cue weightings is needed for each of the tasks of interest. A weighting system, such as that aforementioned, can be employed to prioritize first, the sensory modality, and then the cues within the modality.

It is here where some simplification can be introduced to the task of quantification. A decision can be made early in the design process to identify what might be called *primary* cues, which are essential to the task and *secondary* cues, which are supportive, but not essential to task completion. Task completion means that the task has been carried out at a level of performance that would be expected of an experienced vehicle operator. For example, there appear to be two primary visual cues for landing an aircraft: One is the angle formed by the runway edges and the second are texture–density cues in the intended landing area of the runway.

Where secondary cue weightings are available, cues weighted below 0.10 or 10% of the total weight of the available cues should be considered secondary cues. One such example is the relative size cue for distance for objects less than 10 m from the observer. In the case of the relative density distance cue in Table 6.1, the cue would be considered only supportive at all distances studied. This is due to its weak relative weighting when compared to other available cues. Note that cue weightings are always in reference to other cues that are available for the particular task. If some of these cues are absent for any reason, the weightings of the remaining cues will necessarily change. The relative density cue might become a primary cue if the particular visual scene is largely barren of other distance cues.

Whether or not the absence of a primary cue from a design specification would necessarily disqualify the device from the standpoint of acceptable perceptual fidelity would depend on the importance of the particular task in assessing the overall value of the device. This is a decision that needs to be made by the customer and not the designer.

A secondary cue does not place as much demand on the design specification. For example, lateral acceleration cues are known to have some effect on aircraft landing performance under some conditions. The same inertial cues also have an effect on the driver performance in the cases of rapid acceleration or deceleration and in maneuvering around sharp curves. However, these inertial effects tend to be very small relative to the primary visual cues in most of the studies of motion platform cueing effects.

In these examples, the absence of secondary inertial cues does not mean that the task cannot be performed, only that it cannot be performed at the absolute maximum level of operator capability. This may not matter in training and testing though it may matter a great deal in the design of advanced research simulators in which small differences in cueing effects may be of importance to a particular control theory or to vehicle system modeling. In any case, dividing perceptual cues in to primary and secondary categories is intended to give the designer a quick view into what will be necessarily in the design specification and what may well be left out and how this division of perceptual cueing might affect the ultimate cost and benefit of a simulation device.

CALCULATING THE INDEX OF PERCEPTUAL FIDELITY

Identifying primary and secondary cues that support a task is a useful preliminary step to a comprehensive approach to the definition of perceptual fidelity. However, a detailed analysis is still required of the perceptual cues for individual tasks to be performed in order to determine the perceptual fidelity of a device. Ideally, individual tasks should be available in the requirements document for the system. If not, the designer will need to define the tasks in consultation with the customer. In any case, the tasks need to be defined at a level of detail that allows for an unambiguous assessment of the supporting perceptual cues. For example, a driving simulator for passenger car use would support behaviors such as speed control, maintenance of lane position, safe vehicle following, passing maneuver, intersection turns, and so on. Note that these tasks are all defined at the level of vehicle control and closely tied to sensory and perceptual processes. Tasks that are primarily cognitive, that is, they involve primarily cognitive component processes such as decision-making or problem-solving that is too remote from perceptual processes and human–vehicle interaction to benefit from design specifications defined by perceptual fidelity. Separate analyses specialized for these tasks are available. Cognitive task analysis (CTA) that decomposes higher level tasks and task goal hierarchies is one such tool.

Perceptual fidelity in vehicle simulation is intimately connected to the issue of vehicle control because the use of perceptual cues is fundamental to control of a real-world vehicle. With the exception of aircraft flown solely with reference to instruments within the cockpit, all vehicles rely heavily if not exclusively on perceptual cues from the external environment. This means that there are essentially two distinct categories of perceptual tasks performed in a vehicle. The first category includes perceptual cues that inform the operator with regard to the state of the vehicle at any given time. They include the ones listed earlier such as those involved in control of the speed of the vehicle. They are derived from the vehicle itself such as vibration and inertial cues and from the external environment such as optic flow and aural cues of motion. The second category of perceptual cues is specifically related to control actions and control device. These include the forces applied to and fed back from the steering wheel and displacement of foot controls such as the accelerator and brake pedals.

Vehicle control and, therefore, the simulation of vehicle control can be described as a collection of tasks which, from the perspective of the vehicle operator, are a combination of perceptual judgments and vehicle control actions. A constant feed forward closed control outer loop action based on the desired and perceived vehicle state (e.g., vehicle speed) and an inner loop dependent on the desired and perceived state of control devices (e.g., steering wheel) is described.

Task Requirements and Decomposition

An index of the perceptual fidelity of a vehicle simulator is constructed by first defining the tasks that the vehicle operator is expected to carry out in the device. This is usually conducted with input from the device customer or a related source. The task requirements document identifies every task for which the device is designed to support. Each of the tasks must be described in sufficient detail that will allow an analyst to decompose the tasks into supporting perceptual cues. A task requirement document must be specific not only with regard to what task the vehicle operator performs but also the expected level of performance of the operator as well (performance criterion). Operator performance levels should be based on a level of performance representative of the intended user population.

Once the task requirements document is completed, the process of task decomposition begins. Typical task analyses that are used by analysts to decompose tasks are often too vague or superficial to be of use in developing an index of perceptual fidelity. This is likely due to the absence of expertise in identifying perceptual elements as task analyses generally rely on subject matter experts whose expertise is in the operation of the vehicle and not in perception. The analysis of the required tasks needs to be conducted by analysts with a high level of expertise in sensory and perceptual processes as well as the perceptual limitations and capabilities of the typical vehicle operator. In this case, the subject matter expert plays a supporting role.

The analysis continues from the task requirements document to a listing of the perceptual cues for each task. In general, most vehicle simulator customers have an expectation that the device design will support a variety of tasks for which the real vehicle was originally designed. Decomposition of each of these tasks into task goals and task subgoals as well as their supporting perceptual cues is then conducted.

For a passenger vehicle or truck, tasks such as maintaining speed within certain limits, safe vehicle following, safe passing of other vehicles with oncoming traffic, intersection turns, and others are likely to be on the task requirements list. Each of these tasks requires a certain level of perceptual fidelity in order to produce a level of driver behavior that is comparable to that of drivers in real-world operations. The details of the decomposition process will be described in Chapter 7. At this point, the focus will be on constructing the perceptual fidelity index (PFI) for each individual task.

Driving tasks that are primarily perceptual such as speed control, lane position, and vehicle following can be decomposed into the cues which support them. In the case of speed perception, research has shown that the perception of self-motion relies heavily on visual cues such as optic flow.

The rate of change in the relative sizes of objects in the visual scene is yet another cue to speed; the higher the vehicle speed, the greater the rate of change in the perceived size of objects within the driver's field of view. A second cue used in speed control is the longitudinal inertial cue, which occurs during acceleration or deceleration. Lateral inertial cues can also occur during maneuvering around sharp curves. Vibrations due to road surface imperfections are positively correlated with speed. Audio cues associated with engine rpm and wind noise are also potential cues to vehicle speed. Certain somatosensory cues are associated with rapid decelerations in vehicle speed due to pressure from seat and shoulder belts on the upper torso.

Calculating Perceptual Fidelity Index for an Individual Task

The simulator design requirements for a system that attempts to achieve perceptual fidelity will contain two versions of the task-level PFI. One version, the desired PFI or PFI_{des}, is an index that is calculated based on known and empirically established estimates of the effect of one or more perceptual cues on the execution of a particular task. The second version, the obtained PFI or PFI_{obt}, is derived from the actual design specifications for the device. The ratio of PFI_{obt} to the desired PFI_{des} is the degree to which the design specification achieves the desired level of perceptual fidelity.

In order to calculate a PFI_{des} for an individual task, the degree of importance of each cue in supporting the task needs to be determined. As control of vehicle speed is critical to vehicle operation, an example emphasizing the contribution of perceptual fidelity to speed control is appropriate. (Translating cues into specific design specifications will be addressed in Chapter 7).

The PFI_{des} of the speed control task consists of the sum of the variance estimate for all of the cues that support the task. The optic flow cue accounts for the largest variance in the speed control task. The most reliable estimate of the contribution of the optic flow cue will be derived from the research literature in which the amount of optic flow is systematically varied and its effects on vehicle speed control are assessed. The variance estimate can be derived through various methods, though here the use of the effect size, R^2, is recommended. Using the speed production data of Diels and Parkes (2010), the effect size cue contribution of optic flow is $R^2 = 0.31$. In other words, 31% of the variance in speed control appears to be due to optic flow.

Additional cue components of speed control known to have an effect on speed control are tactile cues associated with lateral accelerations in turns and the vibrations through the seat pan from road disturbances (deGroot et al., 2011). The former cue is associated with inertial forces that push the driver into the side of the seat in the direction opposite to that of the turn. The latter cue, seat pan vibrations, is a result of tire contact with irregular road surfaces and will increase with an increase in vehicle speed. In the study by deGroot et al. the effect size of the lateral acceleration cue on speed control was $R^2 = 0.05$, whereas the effect size for the seat pan vibration cue was $R^2 = 0.11$. The sum of the weights for this task will be the sum of variance estimates or $R^2 = 0.47$. The sum of the variance estimates for this task is best estimate of the PFI_{des}. That is, the PFI_{des} is always determined by the empirically derived variance estimates for each cue that supports the particular task.

Each subsequent task from the task requirements document is subjected to the same calculation of estimating how much each cue in the task accounts for what amount of variance in the behavioral measure. For example, the next task in the document might be emergency braking. In this task, as in the first, all of the known cues that contribute to this task are quantified from the research literature using the measure of effect size (R^2), once the individual cue weights are identified and summed to derive the PFI_{des}.

The process continues through the task requirements document until a PFI_{des} is calculated for each task. Note that the desired level of perceptual fidelity for each task is based on the available evidence at the time the design specifications are written. This means that, in some cases, data on the strength of a particular cue or set of cues for a specific task will be unavailable or incomplete. This fact should be noted in the design specifications document.

At the completion of the task level calculations for PFI_{des}, the analyst may wish to begin the process of determining what design components are possible given the budget provided. For each task, a calculation can then be made of how each task would be affected by the elimination of a design component and the perceptual cues it provides. The analyst would then recalculate the sum of the cue weights for each task affected by the elimination of the design component. The sum of these corrected cue weights for each task is the PFI obtained (PFI_{obt}). The ratio of the PFI_{obt} to the PFI_{des} or PFI_{obt}/PFI_{des} will provide the analyst with a measure of the impact of the elimination of that particular design component on each task. In the example on cue requirements for speed control aforementioned, elimination of the tactile cues provided by the driver seat would yield a PFI_{obt} of 0.34 and a PFI_{obt}/PFI_{des} of 0.34/0.47 or 0.78. Typically, this ratio would be expressed as a percentage in which the particular set of design specifications would achieve 78% of the desired level of perceptual fidelity (PFI_{des}) for this particular task.

A General Index of Perceptual Fidelity

A general index of perceptual fidelity or PFI of a vehicle simulator or any virtual environment is the overall ratio or PFI_{obt}/PFI_{des} for all tasks that the device is intended to support. Thus, the overall perceptual fidelity index is the sum of the PFI_{obt} (ΣPFI_{obt}) for all tasks divided by the sum of the PFI_{des} (ΣPFI_{des}) for all tasks.

The resulting ratio, $\Sigma PFI_{obt}/\Sigma PFI_{des}$, quantifies the degree to which a particular design specification supports perceptual fidelity. Generally expressed as a percentage, the ratio quantifies the degree to which the overall design specification meets the desired level of perceptual fidelity.

Perceptual fidelity is defined here, not as a measure of the degree to which a device matches that of its physical counterpart, but the degree to which it provides the necessary perceptual cues for each task that the operator needs to accomplish with the device. If the design specifications for the device provide all the perceptual cues needed to support all of the tasks, then the maximum achievable level of perceptual fidelity will be obtained. In most cases, however, it is likely that this level will not be achieved by most designs due to a variety of factors. The most common factor will be cost, particularly with regard to the provision of inertial cues, which may require expensive motion platforms. However, the use of the PFI will aid the designer in determining which cues will provide the most benefit for a given level of investment.

Other Means of Quantification

Calculating PFI measures using empirically derived effect sizes from real or simulated vehicle control studies is likely to provide the most reliable estimates of how perceptual cues affect the operator behavior. These studies measure the actual vehicle control behavior in either actual or very similar conditions to those in the real world.

However, effect sizes for all of the perceptual cues that might affect the operator behavior may not be available. In this case, alternative means of estimating the effect of perceptual cues on behavior are needed. Alternatives such as those using Bayesian estimates such as the JND^2, and the estimate of cue reliability derived from it, $1/JND^2$, can be used. Although these estimates may be useful in calculating estimates of cue effectiveness, they usually do not involve vehicle control behavior as the measure of behavior. Most of these estimates are derived from forced-choice responses. These forced-choice responses can provide useful estimates of the cue contribution to basic perceptual behavior such as distance judgments, perception of motion, or force feedback perception. As they do not include control behavior, they cannot measure either the relationship between perception and control nor can they provide assessments of how control device cues such as force feedback affect control behavior.

SUMMARY

This chapter began with the identification of different means by which perceptual fidelity could be measured. One means of defining perceptual fidelity is based on the basic operating parameters of the sense receptor. This sense-limited or sensory fidelity is measured by the degree to which a vehicle simulator provides sensations which meet or exceed the minimal stimulus energy necessary to trigger a sense receptor. The second measure of perceptual fidelity is based on higher level perceptions such as the ability to judge distances or to discriminate differences in the degree of self-motion. A review of modern perceptual theory is discussed with emphasis on the Bayesian view of perception. A description of how to calculate Bayesian weightings of perceptual cues was provided as well as examples of how to use them in design. The use of effect sizes of cues is described as is their use in quantifying an index of PFI. The use of PFIs in design and how they can be used to reveal the effects of design components in individual tasks and in an overall evaluation of a device are discussed.

REFERENCES

Beukers, J.T., Stroosma, O., Pool, D.M., Mulder, M., and van Passen, M.M. 2009. Investigation into pilot perception and control during decrab maneuvers in simulated flight. *AIAA Modeling and Simulation Technologies Conference*, August 10–13. Chicago, IL.

Bowden, J.L. and McNulty, P.A. 1972. Age-related changes in cutaneous sensations in the healthy human hand. *Age*, 35, 1077–1089.

Brunswik, E. 1955. Representative design and probabilistic theory in a functional psychology. *Psychological Review*, 62, 193–217.

Clark, B. and Stewart, J. 1968. Comparison of methods to determine thresholds for perception of angular acceleration. *The American Journal of Psychology*, 81, 207–216.

Cutting, J.E. and Vishton, P.M. 1995. Perceiving layout and knowing distances and knowing distances: The integration, relative potency, and contextual use of different information about depth. In *Perception of Space and Motion*. New York: Academic Press.

deGroot, S., deWinter, J.C.F., Mulder, M., and Wieringa, P.A. 2011. Nonvestibular motion cueing in a fixed-base driving simulator: Effects on driver braking and cornering performance. *Presence*, 20, 117–142.

deWinter, J.C.F., Dodou, D., and Mulder, M. 2012. Training effectiveness of whole body flight simulator motion: A comprehensive meta-analysis. *The International Journal of Aviation Psychology*, 22, 164–183.

Diels, C. and Parkes, A.M. 2010. Geometric field of view manipulations affect perceived speed in driving simulators. *Advances in Transportation Studies*, 22, 53–64.

Green, D.M. and Swets, J.A. 1988. *Signal Detection Theory and Psychophysics* (rev. ed.). Los Altos, CA: Peninsula Publishing.

Gundry, A.J. 1978. Thresholds of perception for periodic linear motion. *Aviation, Space, and Environmental Medicine*, 49, 679–686.

Lee, A.T. and Bussolari, S. 1989. Flight simulator motion and air transport training. *Aviation, Space, and Environmental Medicine*, 60, 136–140.

Lovenberg, J. and Johansson, R.S. 1984. Regional differences and interindividual variability and sensitivity to vibration in the glabrous skin of the human hand. *Brain Research*, 301, 65–72.

Palmqvist, L. 2013. Depth perception in driving simulators. Bachelors thesis, University of Umea, Umea, Sweden.

Ricard, G.L., Parrish, R.V., Ashworth, B.R., and Wells, M.D. 1981. The effects of various fidelity factors on simulated helicopter hover. (No. NAVTRAEQUIPC-IH-321). Orlando, FL: Naval Training Equipment Center.

Zaal, P.M.T., Schroeder, J.A., and Chung, W.W. 2015. *Objective Motion Cueing Criteria Investigation Based on Three Flight Tasks*. London, UK: Royal Aeronautical Society Flight Simulation Conference.

Wang, A.-H. and Chen, M.-T. 2000. Effects of polarity and luminance contrast on visual performance. *International Journal of Industrial Ergonomics*, 25, 415–421.

7 Implications for the Design Process

INTRODUCTION

This book began with a discussion of the problems of fidelity in virtual environments especially those environments in which the goal of the design is to create an environment which will support a desired behavior. In the case of vehicle simulators, the desired behavior would include the ability to manually control the vehicle, for example, with reference to external visual cues. The perceptual fidelity of the visual scene would depend on its ability to provide the critical perceptual cues necessary to carry out the task and not on whether or not the visual scene was photorealistic.

A design that defines the goal of a virtual environment in terms of user behavior must support and must have a design paradigm that is inherently different from one that supports physical fidelity. In the latter case, the design goal is physical replication of the desired system and the environment in which it operates. In the case of vehicles, this includes the physical structure and contents of the vehicle such as control devices and instrumentation. It would also include a photorealistic simulation of the external visual environment as well as the physical forces acting on the vehicle.

The use of behavior-based design paradigms such as perceptual fidelity will have inevitable consequences for the design process of vehicle simulators. How radical these changes are will depend on the approach that is currently being taken in designing for physical replication. In its purest form, physical replication will essentially ignore the issues of importance to perceptual fidelity and concentrate solely on reproducing the appearance and function of the vehicle and its operating environment. In this case, perceptual fidelity with its emphasis on perceptual cueing and behavioral validation will be a dramatic departure from the existing design process. It will require not only a different focus of the design process but also personnel with expertise that is radically different from the hardware and software engineers typically involved in simulator development.

In other cases, where customers or market research make clear demands that the device support specific tasks and performance criteria, the changes in the design process will be much less dramatic. For those already accustomed to task-oriented design and the human-centered design paradigm, implementing a design approach which stresses perceptual rather than physical fidelity should be adopted with relative ease. The following steps summarize the design process for perceptual fidelity. The remainder of the chapter will provide the details for each of these steps.

The goal of achieving perceptual fidelity in a vehicle simulator involves several separate analytical tasks. The first, the task requirements document, includes the identification of all tasks that the vehicle operator will be required to carry out in the device. The document will also include the performance criterion for each of these tasks. The performance criteria for each task will identify any attribute within the device user population that might affect their performance. These include user perceptual limitations, for example, in vision or hearing. The second step in the design process is the task analysis. This is a familiar activity for those accustomed to human-centered design. The version of task analysis used for perceptual fidelity focuses not only on procedural tasks but also on tasks that incorporate perception or perceptual-motor activities as essential components. Typically, tasks are often subdivided into further subtasks such as those that might involve purely perceptual distance judgments and those that have a strict vehicle control component. The third step is the task decomposition process. Tasks or subtasks are decomposed into task elements each of which contain one or more perceptual components and their associated perceptual cues. The fourth step in the design process converts the perceptual cues into estimates of their relative reliability and thus their utility for the vehicle operator. These estimates or weights will typically be derived using the Bayesian methodology with empirically derived measures of perceptual cue discrimination efficacy, although other methods that identify perceptual cue weightings might be used.

The design process ends with the calculation of the perceptual fidelity index (PFI). The development of the PFI, both desired and obtained, allows the designer and customer to identify potential trade-offs between design components and their potential impact on perceptual fidelity. As PFIs are calculated for each task, the potential impact on individual tasks can be readily assessed.

THE DESIGN PROCESS

Up to this point, the discussion of perceptual fidelity has been quite general with regard to the actual design process. In order for the goal of perceptual fidelity to be achieved it must be incorporated into the design process at a very early stage. Ideally, this will occur before any design prototype is developed. Moreover, the resulting design specifications of this process must be accompanied by an estimate of how specific behaviors are likely to vary both within an individual and among individuals in the target population.

Some of these design process stages are similar to what is called user or human-centered design. However, the implementation of the latter has been, in practice, quite vague on specifics often relying heavily on market research data, on cursory walk-through by untrained personnel, by usability testing or by reliance on subjective ratings of usability or user experience.

The design process advocated here is much more rigorous than traditional human-centered design and necessarily requires expertise in the sensory and perceptual processes that is not widely available in the simulator industry. Information on basic sensory and perceptual processes available in this book, however, should aid in the design process. Nonetheless, reliance on published literature in both basic and applied perception will be a necessary element of the design process if high levels of perceptual fidelity are to be achieved.

USER POPULATION LIMITATIONS

Vision

Deciding on the user population for a particular device design is relatively straight-forward. For a commercial product, market demands will determine the specific demography that would be using this particular device. The driving simulator or flight simulators of the user population are restricted to those either in driving or in flight training or those who are already licensed. Licensing necessarily imposes certain restrictions on who is eligible. Basic visual acuity requirements typically include correctable acuity, for example, to a Snellen acuity of 20/40 for drivers and 20/20 for commercial pilots. Although color deficiencies are not generally tested for a driver's license, pilots at all levels of licensing are required to pass basic color discrimination tests as color perception is important for flying especially for night operations. Consequently, color rendering accuracy is more important for flight training that it is for driving simulators.

For devices marketed to the general population, the designer will be faced with a very large variation in visual acuity from near blindness at Snellen 20/200 to 20/20 or better. As many modern vehicle simulators have a relatively poor display resolution of about 3 arc min per pixel, this variation in visual acuity is not yet of great concern. However, with the introduction of ultra high-definition displays, the display resolution can easily match 20/20 or even better, leaving those with poor uncorrected visual acuity unable to discern the potentially critical details in the displayed image. For conventional displays, this may simply mean requiring individuals to wear corrective lenses. However, for HMDs in which the user supplied lenses cannot be accommodated, the device will need to provide the correction itself. As many individuals may have different accommodation requirements for each eye, the device will need to provide user-adjustable lenses for each eye. This is especially true for HMDs which provide stereoscopic images as the intended stereopsis will be attenuated or eliminated by uncorrected acuity differences in the two eyes.

As noted earlier, color perception requirements are driven by the particular environment in which the user operates. For driving environments, color coding in signals and signage is employed for discriminating information that is essential to the driving task and is defined by the transportation authority for the particular locale. Amber, green, and red colors are commonly used in traffic signals, whereas red and amber colors are used for vehicle braking and turn signals. Similarly, blue and red lights differentiate runways from taxiways for pilots and red and green are used for aircraft navigation and anti-collision lights. In the flight regime, quite specific colors are used, so they are routinely specified in regulations controlling the flight simulator design.

Colors are also used in signage such as stop, yield, no entry, and others in driving environments. Typically, the color is used as a redundant cue. In the case of stop signs, the color red is used on a sign that is always octagonal in shape. Similarly, yield signs are colored yellow on a sign that is always triangular in shape. Information signs are usually rectangular with green or blue backgrounds. This signage color strategy varies from country-to-country somewhat but the use of color cue redundancy is fairly common. In addition, traffic light signals are differentiated by color but also by position as the red (stop) light is always on the top and the green (go) light is always on the bottom. The signage design that uses additional cues, such as shape or position, redundant to color allows those with poorer color perception to still function safely as drivers. Driving simulator design needs to maintain this redundant cue strategy for the same reason.

Virtual environments targeted at the general population are more likely to simply require basic color rendering that allows for the use of the standard 24-bit (true) color with which most computer monitors and projection systems are equipped. Color deficiencies occur in about 8% of males and 0.2% of females in the general population, so designers need to be aware of these user limitations. For example, red–green discrimination deficiencies are the most common so that the designer should either avoid using these colors or use deeply saturated colors if red–green discrimination is necessary. In addition, as display luminance affects the color perception as well as acuity, display luminance values at or above 300 cd/m² should be used. In some cases, the context in which the color is presented may alleviate this problem.

Humans respond primarily to the hue in color. As hue is roughly equivalent to wavelength, colors should be displayed in accordance with commission on illumination (CIE) color standards, which are based on wavelength rather than on the color defined by equipment manufacturers in which rendering of a specific color can widely vary. Correct color rendering is especially important for clinical applications in which correct color identification of displayed elements may be essential for diagnosing patient color deficiencies.

Stereopsis

Stereopsis, the depth perception that accompanies binocular disparity in image displays, will also vary in the general population. It is usually not tested in drivers but may be tested in pilots and others in which depth perception is deemed critical. Interpupillary distance (IPD) varies in the general adult population. In a study of the nearly 4000 adults aged between 17 and 51 years, IPD ranged between 52 mm (2.05 in.) and 78 mm (3.07 in.) with an average of 63.4 mm (2.5 in.) (Dodgson, 2004). The data formed a normal distribution with a standard deviation of 3.8 mm (0.15 in.). This meant that 95% of the adults were tested had IPDs within 56 mm (2.2 in.) to 71 mm (2.8 in.).

The accommodation of this IPD variance in the design of HMDs is necessary in order that the display projection is centered correctly on each eye for each of the individual HMD displays. This is particularly important for stereoscopic HMDs intended for use to aid depth perception within very short ranges as is the case, for example, in surgical training applications. At this distance, small variations in binocular disparity can have large effects on behavior.

Even correctly designed HMDs may not be enough to produce stereopsis in all device users. There are deficiencies in the general adult population with respect to the production of stereopsis. Stereo blindness occurs in about 3% of the general population. In this condition, the normal depth cueing properties provided by binocularly disparate images will not be available.

Prescreening of test subjects intended for use in the evaluation of stereoscopic HMDs is recommended.

Myopia, Hyperopia, and Presbyopia

Other conditions of visual impairment will affect users of virtual environments. Approximately one-third of the adult populations of the United States and Western Europe are afflicted with these deficiencies. They include myopia (near vision), hyperopia (far vision), and presbyopia. The latter affects most adults over the age of 50 and is a deficiency of near vision acuity due to hardening of the lens of the eye. In general, these deficiencies are addressed in device designs with built-in optical corrections in the case of HMDs and by allowing the use of corrective lenses in the case of conventional displays. Most HMDs do not have the space available between the eye and the display surface to permit the use of corrective lenses that users might normally wear. If the HMDs do not provide some measure of optical adjustment, the user with these deficiencies might find the HMD to be unusable. As corrections may need to be different for each eye, independent optical adjustments for each eye should be available.

Vestibular

In the general population, vestibular responsiveness is likely to be modified to some extent by disease, genetic abnormalities, and other factors. The most serious of these maladies are those that impact the vestibular system itself. One such disease, Meniere's disease, afflicts about 0.2% of the population (Lopez-Escamez et al., 2015). This disease may result in damage to the vestibular structures resulting in what is known as a labyrinthine defective individual. These individuals are incapable of a vestibular response to physical accelerations. Their entire response to physical movement is due to their perception of the resulting visual or somatosensory stimuli. They are effectively immune to vestibular input and are consequently immune to any motion sickness that might be caused by a virtual environment.

In general, the more common and less serious vestibular problems occur in the adult population over the age of 40. Approximately, 35% of the U.S. population in this age group have some deficiency in vestibular responsiveness that would produce problems of balance and vestibular vertigo (Agrawal et al., 2009).

The only means to assure that the individual user has a normal functioning vestibular system is the use of the caloric test or other clinical tests of vestibular function. This is generally only practical in clinical or research settings. Older adults (>65 years), however, are more prone to vestibular problems than the general adult population. Problems of balance and postural control tend to increase as individuals grow older as a result of vestibular maladies as well as a functional decline in other sensory systems related to balance.

In general, devices dependent on the delivery of physical motion stimuli, such as motion platforms, should expect that some individual users will not respond properly. This is largely a problem for those devices intended for use by the general adult population. These include driving simulators with motion platforms. Motion platform systems designed for use by professional pilots are less likely to encounter users with vestibular problems due to the screening of these pilots for problems in vestibular perception during their career. Notably, the pilot population does not appear to be any more sensitive to vestibular inputs than college age adults

(Clark and Stewart, 1972). Thus, it is not necessary to increase the performance accuracy of motion platform systems beyond that is acceptable for use by young adults.

Somatosensory

Not surprisingly, proprioceptive performance declines with increasing age. The decline in accuracy in positioning of the limbs is a particular issue in operating vehicles. For older adults, the addition of high mental workloads associated with vehicle operation appears to aggravate the decline in proprioceptive performance (Goble et al., 2012). This is generally attributed to the decline in function of peripheral sensory receptors in the limbs. This may account for accidents that involve pedal misidentification in some driving accidents in which older drivers mistake the accelerator for the brake pedal.

Many simulated control devices will require accurate force perception in the fingers and hands for proper operation. In this case, cutaneous mechanoreceptors in the fingers and hand are the principal sensory receptors that are involved. As noted in Chapter 4, cutaneous mechanoreceptor function declines significantly with age. Adding to this decline in performance are a variety of diseases, such as diabetes mellitus, which often result in peripheral neuropathy. These diseases and other maladies result in a rate of peripheral neuropathy ranging from 2.4% in young adults increasing to 8.0% in older adults (Shields, 2010). In these individuals, accuracy in touch of the hands and fingers is difficult and sometimes impossible.

USER PERFORMANCE VARIABILITY

It is essential for the designer to accept the fact that user perceptions and perceptual processes vary to some extent from one individual to another within the device user population. Unlike designs based on physical fidelity, those driven by the goal of perceptual fidelity must incorporate the concept that there is variability in perceptual ability not only from individual-to-individual but within an individual from one time to another. This is particularly important for any behavior-based design goal so that the range of user performances can be incorporated in the design specifications.

In many areas of human performance, the variation in perceptual performance is a means of describing the range, both lower and upper limits, of perceptual and perceptual-motor ability. To describe the ability of an individual user to perceive or to apply a specific force, the range of each of these must be known and expressed in units of force (e.g., Newton). Similarly, to assess the ability of the user to discriminate among different levels of perceived force, a measure of discrimination, such as the just noticeable difference (JND) is provided. Thus, although the design of a control system may specify a level of force under a specified set of conditions, the perception of that force should be expressed as both a single value, a measure of central tendency (e.g., mean or median), and as a measure of variation (e.g., standard deviation or range). The designer then will not be surprised to find during device testing that a test with one individual will result in one value, whereas a test with another individual will result in a slightly different value. The designer needs to be aware, not only of the inherent variation of individuals for this particular performance value, but also the acceptable range of variation. This avoids the mistake of identifying the normal performance variation from one individual to another as an anomaly or error. It will also allow the designer to identify behaviors that are outside the normal expected range and to investigate the reasons for this aberrant behavior.

Estimating the variance of human perceptual performance becomes more complex, but not necessarily less accurate, the further perceptual processes are from basic sensory processes. This is due to the influence of intervening structures in the nervous system and, more important, to the influence of variations in experience as more elements of higher level cortical systems are involved in the process.

Specialized User Populations

Identifying the user population of the device to be designed is typically the responsibility of those in market research for commercial products. For other cases, such as flight or driving simulators intended for professional use, an understanding of the special capabilities of the user population is required. Attempting to use the same user population parameters of the general commercial audience is likely to underestimate the specialized user population capabilities. Furthermore, design parameters that might be acceptable to the general population are likely to be unacceptable to this specialized group. The demand for higher levels of perceptual fidelity in the presentation of cues affecting vehicle control in a specialized group of vehicle operators is more likely because their skill levels are much higher and the demands on their skills are much greater. This is true of perceptual cues that inform the operator of the vehicle state as well as force feedback cues from control devices. Developing devices for these specialized groups should, where possible, use perceptual and performance data collected from a representative sample of this group. For example, it is generally preferable to use performance data collected from the experienced pilot population in determining the design of a flight simulator visual imaging system than to rely on that collected from the nonpilot population. Similarly, devices designed for use by trained police officers for training in high speed pursuit should rely on data from that population of users as these users are accustomed to controlling the vehicle at speeds far above that of the typical driver.

TASK ANALYSIS

In Chapter 6, the importance of understanding and describing the task required of a device was emphasized. The task requirements document is a means of guiding subsequent analysis of the perceptual fidelity of a device. It is therefore critical to the success of a device design that a detailed description is developed of the tasks that users will perform in the device. The level of specificity of the tasks must be sufficient to allow subsequent identification of the perceptual cues necessary to support these tasks.

The task requirements document has to provide task definitions, which are quite specific regarding what is and what is not a part of the task. Generic behavior such as speed control in a passenger car or truck should be explicitly defined and should include all task conditions which might affect speed control behavior such as road curvature, slope and road condition, emergency braking, intersection turns, and so on. The statements regarding the control devices that would be expected to support vehicle speed control such as the use of brakes, accelerator pedal, gear shift, and clutch also should be included. All of these may have unique cueing properties that may be used by the operator.

Task Definition

Task analyses are procedures by which the tasks that the vehicle operator performs can be explicitly identified and decomposed into their constituent elements. A variety of techniques exist for conducting task analyses (Kirwan and Ainsworth, 1992). But some general rules apply to all forms of task analysis. First, tasks have definable goals and objectives. Execution of a task or a sequence of tasks begins with the identification of what that task or sequence of tasks is to accomplish. For a vehicle operator, the task may be as simple as turning a switch or as complex as landing an aircraft. Second, every task goal or objective has a criterion or criteria that inform the vehicle operator that the task has been satisfactorily completed. The turning of a switch has a performance criterion defined by its position. Once the position is reached, the task goal or objective has been achieved. The criteria for landing an aircraft are reached when all of its wheels are on the runway and aircraft speed has slowed to a point where further flight is no longer possible.

As this book is concerned with perception and related behaviors, the task of interest are those which rely on perception in order for them to be completed. Each such task contains within it perceptual and perceptual-motor components that rely on input from the vehicle and in some cases from the surrounding environment. Correct identification of these perceptual sources of information or perceptual cues can be divided into information regarding the state of the vehicle on the one hand, and the state of the device that controls the vehicle on the other.

In the aforementioned example, the task of positioning a switch is usually driven by the visual feedback regarding the specific position of the switch. Perceptual cues may also include tactile or proprioceptive force feedback to the hand from the physical position of the switch. Even the simple task of operating a switch is usually accompanied by feedback from other sources of information both inside and outside the vehicle. A simple on–off switch for vehicle headlights may be accompanied by an auditory feedback from the switch itself as well as indicator lights on the vehicle instrument panel. Feedback from headlight reflections on objects outside the vehicle is an additional cue to the state of the control device.

In the case of aircraft landing, a variety of cues are available to the pilot in determining whether the landing has been achieved. Although vision plays an important role in landing, the cues to the contact of the wheels on the runway are mostly the physical forces that are transmitted through the aircraft. This contact force can be felt through the seat pan and sometimes through control devices such as the control yoke or control stick. Although external visual cues are essential to establish a correct landing profile and initiation of the prelanding flare, they can only give an approximation of the height of the wheels above the runway.

Actual contact with the runway surfaces is often accompanied by auditory cues as well as physical ones. These include those associated with tire scrubbing on the surface of the runway as well as sounds from the contraction of the landing gear system as it takes on the weight of the aircraft.

The visual system plays an essential role in maintaining the aircraft on the runway during the rollout after contact has been made with the runway. This is particularly true for maintaining aircraft position on the center of the runway and avoiding drifting to either side. Although visual cues from the external environment give the pilot a sense of approximate ground speed, the airspeed indicator inside the cockpit is more accurate with regard to the actual speed.

These two brief examples of tasks have identified each of the task goal or objective, the criterion for successful task completion, and the perceptual cues that are needed to achieve that criterion. For the simple case of switch positioning, task definition is relatively straightforward because the goals and actions are readily definable. In the case of more complex behaviors, such as landing an aircraft, task definition is more difficult. The landing of an aircraft is really just the endpoint of a long series of actions that begin with the initial approach to the runway, followed by the final approach, the initiation of prelanding flare, the landing itself, and finally the roll out of the aircraft on the runway. Failure to execute any one of these cases of approach and landing will have a negative effect on the ultimate outcome. Thus, for many complex tasks there are often many components that are needed to be defined separately.

As this book is concerned with perception, decomposition of a task should end when the perceptual cues supporting the task are fully defined. If the task is simply one of a series of interrelated tasks, then the delineation of one task from another is determined when the task goals or objectives have changed. In any case, the task analysis is not complete until all of the perceptual cues supporting the vehicle operator's perception of the vehicle state and the state of the control devices are described for all of the tasks in the task requirements document.

Task Decomposition

The task requirements document would include summaries of the sort just described as a means of defining what task will be carried out in the vehicle simulator and the acceptable criteria for performing these tasks. However, the information necessary to support perceptual fidelity in vehicle simulation requires a finer level of detail. The decomposition of the task elements into smaller units is needed to reveal the perceptual cues that support each element of a task.

A variety of decomposition strategies exist but most have evolved from the original work of Miller (1953). The details of this work are described in Kirwan and Ainsworth (1992). Briefly, decomposition of a task reduces it into constituent task elements or individual actions. Seemingly simple tasks, such as turning a switch, are decomposed into ever finer task elements. For the purpose of perceptual fidelity, the decomposition of the task is complete when all of the sensory and perceptual components have been identified. Modification of the basic decomposition categories used by Miller has been done here in order to elucidate the perceptual components for each task element. The reader may wish to include additional decomposition categories for other reasons, but here we will concentrate on the perceptual task and perceptual cues.

The decomposition categories for perceptual fidelity described here are not all inclusive for every conceivable vehicle. The reader is encouraged to modify the list as needed for their particular needs. The most important criteria to be followed is that decomposition categories reveal all the perceptual cues that will be needed to support each element of a task.

Decomposition Categories

The first and essential decomposition category is that of task elements. The task elements category includes all individual actions that make up a task. In Table 7.1, a decomposition of the driving task of turning left in front of oncoming traffic (the equivalent of a right turn into traffic for right-hand drive vehicles) is shown. The title of the decomposition table states the task and the task goal or objective in general terms. The task element column in the table lists all of the constituent actions of this task in the order that they would be normally executed to reach the specified task goal. Some task elements may be executed simultaneously even though they are listed in sequence. The sequencing in this case is simply to identify the separate task elements with individual perceptual cues that need to be identified. The simplest way to identify a task element in the decomposition of a task is that it always contains a verb such as *turn* or *accelerate*. As vehicle actions are described, the task elements will typically include some aspect of vehicle operation, including the operational environment such as other vehicles, signage, and road conditions.

The second decomposition column in Table 7.1 identifies the particular sensory or perceptual system involved in the task element. The basic senses such as vision, proprioception, vestibular, tactile, and force perception are included here. All sensory systems that would likely be involved in the performance of the task elements should be identified. These include sensory systems involved in all aspects of vehicle control, including control devices, such as a steering wheel, and all aspects of input from the vehicle and its operating environment. Identifying perceptual cues associated with the use of control devices is important especially for those control devices that rely on nonvisual cues for their operation.

The third column in Table 7.1 identifies a particular perceptual task that is associated with the task element. Most perceptual tasks in driving are visual in nature. They include such tasks as distance and speed judgment. In cases where the driver's LOS is other than that normally aligned with the vehicle heading, this should be identified here with the phrase *LOS off-axis*. This will help to identify the field of view (FOV) for the vehicle simulator design. The first task element, 1.1, reveals that the driver will need to look at the road that is intersecting the road on which the driver's vehicle is traveling. Calculations of the predicted LOS off-axis can wait until the second phase of the analysis, the design phase, in which design component details are defined.

The fourth column in Table 7.1 identifies the perceptual cues that are associated with the perceptual tasks for each task element. The perceptual cue or cues such as optic flow support the perception of self-motion which, in turn, informs the vehicle operator with respect to vehicle speed. Perceptual cues in this column are deliberately kept separate from those associated with control device use so that separate estimates of their respective contributions can be made.

Columns 2, 3, and 4 in Table 7.1 refer to perception as it relates to the state of the vehicle and the environment surrounding the vehicle. Perceptual cues can arise

TABLE 7.1

Task Decomposition for Intersection Turn in Front of an Oncoming Traffic. Clearance of All Oncoming Lanes Must Occur with a Sufficient Margin of Safety

Task Elements	Perceptual System	Perceptual Task	Perceptual Cues	Control Device	Display Device
1.1 Crossing distance to intersecting road	Visual	Distance LOS off-axis	Road texture linear perspective object size		
1.2 Space available on new road	Visual	Area size judgment	Object size		
1.3 TTC oncoming traffic	Visual	TTC	Rate of angular size change of object		
1.4 Current vehicle state	Visual auditory	Self-motion	Optic flow		
1.5 Accel vehicle state	Visual vestibular	Self-motion	Optic flow linear accel	Accel pedal/tactile Proprioception	
1.6 Initiate turn	Visual vestibular	Self-motion heading	Optic flow angular accel	Steering wheel/force	
1.7 Complete turn	Visual vestibular	Self-motion heading	Optic flow angular accel	Steering wheel/force	
1.8 Move vehicle to clear lane	Visual	Self-motion distance traveled	Optic flow distance		Turn signal/auditory

from the vehicle itself transmitted through components of the vehicle chassis is as is the case with vibrations associated with vehicle movement. Perceptual cues from outside of the vehicle and the operating environment will include cues such as distance, speed, position, and others that the operator will extract from that environment. Physical motion cues such as those due to vehicle velocity or inertial cues due to centrifugal forces are included here.

The fifth column identifies the control devices, if any, involving the task element. Control device use such as steering wheel and accelerator and brake operations are generally nonvisual or have associated nonvisual cues at the point in which the control is initiated. Force feedback cues will accompany steering wheel inputs and other input cues. This column should be used to identify the cues associated with the operation of these control devices. Identifying the control cues that are part of the closed-loop control system are particularly important in establishing perceptual fidelity for both maneuvering and disturbance control activities.

The final column in Table 7.1 identifies the display devices that may be involved in the task element. Instrument displays such as speedometer, tachometer, or others are often a normal part of the operation of vehicles. Although not necessarily a part of the perceptual fidelity of a device, some aspects of display devices may be involved in operator perceptual processes. For example, the auditory feedback associated with turn signal indicators is often used by the operator to indicate that the turn signals have been initiated, whereas the operator is visually focused on the oncoming traffic. Other alerting systems associated with vehicle operations are becoming increasingly common as more automation is introduced into vehicle cockpits. Design decisions regarding the perceptual fidelity needed for a vehicle simulator will need to consider including these information display systems in the task analysis and task decomposition process.

Prioritizing Task Elements

The decomposition of a task such as that in Table 7.1 allows the analyst to identify the perceptual task and perceptual cues that are essential to achieving high levels of perceptual fidelity. Inevitably, some tasks and their perceptual cues are likely to be more important than others. In the case of the driver's task of turning in front of oncoming traffic, safe completion of this task is dependent on the ability of the driver to accurately assess the time to collision (TTC). TTC is calculated by dividing the distance of the oncoming vehicle by the rate of reduction in that distance. The latter is solely a function of the speed of the oncoming vehicle when the driver's own vehicle is stationary. This distance is often referred to as *gap acceptance* when the driver has to estimate the time available to clear the opposite lane. This task is further complicated when more than one lane is involved due to the fact that oncoming traffic may be staggered in the opposing lanes. The typical gap acceptance time for passenger cars is between 4 and 7 s. The time will vary depending on a number of factors, including traffic conditions, whether the driver's vehicle is fully stopped, the driver's desired margin of safety, and others.

The second critical task element is the driver's ability to judge the time it will take to cross the lane or lanes. The crossing time must allow for complete clearance of the driver's own vehicle from the path of oncoming traffic. The time includes vehicle acceleration time and the time to execute the turn as well as the lane clearance time.

Intersection turns of this type are a common cause of accidents due to the failure to accurately judge TTC and to judge the time it will take to completely clear the opposing lane and avoid a collision.

The perceptual cues in task elements 1.1, 1.2, and 1.3 might be considered primary cues with regard to the left turn task. Accurate rendering of perceptual cues with regard to the change in angular size of an oncoming vehicle or *object looming* should be given priority. In addition to the primary cues associated with these task elements, the first task element (1.1) has been flagged with off-axis viewing meaning that the driver will need to shift the LOS from that aligned with the vehicle heading or a *straight ahead* LOS to a LOS that requires head or eye movements to either side of the vehicle heading. This has implications for changes in the size of the simulator horizontal FOV to accommodate this particular task.

Finally, control device primary cues can also be identified if they are deemed to be critical to successful task completion. Failure to provide accurate control device feedback may mean that the operator of the vehicle simulator will have to alter his or her control behavior to accommodate the lack of correct device feedback. Accurate force feedback cueing in control operation is vital for tasks such as the left turn in front of oncoming traffic as the driver will rely heavily on nonvisual control device cues in a task dominated by the processing of visual cues outside of the vehicle. Control device perceptual fidelity should be considered as comparable in importance to that of other design components as it is a vital element in the process of vehicle control.

The above-mentioned example of task decomposition is one of the many that will be done with any vehicle simulator system as a means of identifying perceptual cues required to support task elements. All of the task elements required of each of the tasks defined in the task requirements document will need to be subjected to the task decomposition process in order that all of the perceptual cues required of the system are identified.

DESIGN SPECIFICATIONS

Once the task decomposition process has been completed, the design specification process begins. Before the perceptual cues that support each task can be realized in the actual vehicle simulator, they must be translated into terms meaningful to simulator design engineers. This will entail the construction of design statements that are quantifiable and convertible into hardware and software specifications.

To simplify the process of translation of perceptual cues into design specifications, it is prudent to separate those perceptual cues of a general relevance to the state of the vehicle and its environment from those of the control and display devices. Second, perceptual cues that are primary in nature, that is, those essential to task completion, should be identified and their impact on the design should be determined as early as possible in the design process.

Visual Imagery

The first primary cue encountered in the task decomposition example was the driver's LOS change from directly ahead of the vehicle to the area to the left of the vehicle where the intersecting road appears. In order to make a left turn onto the intersecting road, the driver would need to view the road at an approximate angle of 90° to the driver's vehicle heading. With the design eye point midway between the driver's two eyes in monoscopic displays, the vehicle simulator visual imagery system needs to provide horizontal field of view coverage of at least 90° to the left of the vehicle in order to accommodate the left turn task. If the base FOV is ±30° to either side of the design eye point to cover the drivers FOV ahead, an additional 60° of horizontal FOV would be needed. Without the additional FOV coverage, the driver of the vehicle simulator would have to execute the left turn *in the blind*. As the driver needs to estimate the total area of entry available in the intersecting road (Task Element 1.2), the decision to turn has to be made before the turn is initiated and not after the turn.

Two perceptual cues are necessary to complete this element of the task; one will support the judgment of distance to cross the opposing lane(s) and the other will support the judgment of the area available on the intersecting road. In the case of the distance judgment, relative size and height in the visual field are the primary cues for the distances involved. The relative size cue applies to the objects that are in the area, including other vehicles, pedestrians, and signage. Accurate rendering of the object retinal size for all objects at this distance is required to support the distance judgment.

The target area on the intersecting road needs to be judged by the driver to be of sufficient size as to allow room for the driver's vehicle. The intersecting road lanes, for example, may contain other vehicles, which may prevent the use of the target area or the target area may be deemed too narrow to accommodate the driver's vehicle safely. If no vehicle or other objects are in the target area, that target area of perceptual judgment will be based on linear perspective cues from the intersecting road edges combined with texture–density cues from the road surface. If using cue weightings of Table 6.1, the texture–density cue for the 10–30 m distances becomes the predominant cue to distance when height in the visual field and relative size cues are absent. Display resolution and texturing capabilities of the visual imagery system need to be sufficiently robust in order to render these distance cues accurately.

The next task element, 1.3, includes the critical task of estimating the TTC for oncoming traffic. The distance at which that TTC time estimate begins is dependent on the distance of the oncoming traffic from the driver's vehicle. The driver then estimates the TTC based on the closure rate of the oncoming vehicle. This closure rate or rate of change in the distance to an object is adjusted continuously as the object approaches. During this period, the driver makes a time estimate as to when the oncoming vehicle will arrive at a point where a left turn into traffic can no longer be safely executed.

The TTC estimate of the driver is based on the perception of object *looming* or the rate of change in the angular size of the object. Object relative size is a primary perceptual cue in this task. The relative size of the oncoming vehicle plays the most important role in terms of the weighting of cues at distances of around 100 m. For oncoming vehicles at a closer range of 30 m, height in the visual field and relative size of the object may also contribute as cues to distance. At this time, however, the contribution of the rate of change in the height in the visual field of an object to the overall TTC calculation is not known. Object looming properties of objects are affected by display resolution as well as the update rate of the image. Both of these display imaging properties will affect the perception of an object looming.

The task elements for the left turn task, task elements 1.4–1.8, are all vehicle control tasks. Task element 1.4, that is, current vehicle state perception (vehicle movement), is needed in order for the driver to accurately estimate the total turn time required to transit the opposite lane safely. With the vehicle in motion, crossing time will be significantly less than if the driver's vehicle is stopped. Current vehicle motion state is largely perceived from the rate of optic flow. All the display imagery resolution, image contrast levels, image resolution and rendering, image update rate, and image FOV affect the degree to which optic flow cues will be accurately displayed in the simulator.

The remaining task elements accelerate the vehicle, change heading to align the vehicle with the intersecting road, and continue the crossing maneuver until the vehicle is clear of the opposing lane(s). All of these elements require the driver to accurately estimate self-motion as well as the heading changes of the vehicle. Optic flow perception is essential to both of these. The left turn into oncoming traffic is one of the more critical tasks that vehicle drivers perform. The task relies very heavily on the accurate perception of distance and speed. All the factors that affect the visual display of optic flow stated earlier apply here as well.

Force Feel and Feedback

The vehicle control task elements include the three principal control devices in passenger cars and trucks: (1) steering wheel, (2) accelerator, and (3) brake pedals for automatic transmissions with the addition of clutch pedal and gear shift for manual transmissions. Control device operation must rely solely on nonvisual tactile and proprioceptive cues. Steering wheel angle inputs to initiate the turn must be done without the need to look at the steering wheel. Movement of the foot from brake to accelerator pedal depends solely on proprioception of foot position and the accurate pedal pressure required. Correct force feedback from the accelerator pedal assures the driver that correct pressure has been applied. Accurate force feel and feedback from control devices must be accurately simulated.

In the case of manual transmissions, drivers normally do not look at the gearshift to operate it. Rather, they rely on proprioception to guide the hand to the gearshift lever and to select the correct gear position. Looking at the gearshift device during the left turn task is not an option than any driver should be required to exercise. The biomechanical architecture of the simulator needs to accurately reproduce that of the real vehicle in order that drive perception of gearshift position can be done correctly without resorting to vision.

Nonvisual Motion Cueing

A potential for vestibular cueing exists at the very beginning of the left turn, during the turn itself, and when the vehicle is aligned with the intersecting road. However, this will likely depend on how aggressive the driver is in making the turn. Using a 0.1 m/s^2 threshold for linear acceleration and $0.3°/\text{s}^2$ threshold for angular acceleration (Benson, 1990), the provision of this cueing would only be considered if the driver is likely to be exposed to such accelerations for a period long enough for them to be sensed (>0.5 s duration).

An additional case can be made for the provision of auditory cues in the vehicle simulator, for example, in the case of auditory cues associated with engine rpm increase during the acceleration phase of the turn and for the auditory feedback confirming turn signal operation. The provision of these cues is very inexpensive and they are recommended for inclusion in the design. No further analysis will be done for these cues here.

CONVERTING PERCEPTUAL CUES TO WEIGHTS

Once the process of decomposition is complete and the perceptual cues needed to support each task are identified, the process of quantification can begin. This process needs to identify and quantify the perceptual cues that are essential for the completion of task elements. In Table 7.2, the perceptual cues and their associated reliability estimates and their relative cue weightings are shown. The first column lists the perceptual tasks that the driver needs to perform to complete the left turn task. The second column lists the perceptual cues associated with these perceptual tasks. These cues may be from one or more sensory modalities. If more than one modality, the perceptual cue weighting should be provided for both individual modality and multimodality cues. If weights are available only for unimodal cues, it should be noted that weights for the multimodal cueing are not available. The third column lists the JNDs used in the weighting's analyses. The JND column should identify from where on the distribution of JNDs the value has been taken. Most JNDs are taken from the midpoint of distribution, JND_{50}, while here the more conservative JND_{99} is used for the reason given below. In the fourth column, the Bayesian reliability estimate for each cue is calculated ($1/JND^2$). Finally, the relative cue weighting is

---

TABLE 7.2

Perceptual Cue Weightings and Perceptual Fidelity Indexes for the Task of Left Turn in Front of an Oncoming Traffic. No Traffic or Other Objects in the Intersecting Road

Perceptual Task	Perceptual Cue	JND_{99}	Reliability	Relative Weight (PFI_{des})
Subtask A—TTC and Distance to Cross				
TTC oncoming vehicle	Rate of change of angular size	0.09	123	0.74
Distance to cross lanes and clear	Texture density	0.15	44	0.26
Subtask B—Vehicle Control				
Self-motion heading	Optic flow	0.11	83	0.23
Acceleration of vehicle	Linear acceleration	0.14	51	0.14
Steering wheel force perception	Force perception	0.22	21	0.06
Acceleration pedal force	Force perception	0.10	100	0.28
Brake force perception	Force perception	0.10	100	0.28

calculated as follows for each cue: $1/JND^2/\Sigma(1/JND^2)$. The relative cue weighting is the PFI_{des} discussed earlier and is the level of cue contribution desired of the vehicle simulator for this task.

Table 7.2 is the end product of a quantification process that begins with task elements that are deemed critical to the completion of the overall task. The process has been simplified for clarity into two basic subtasks. Subtask A is the combination of the TTC and lane-crossing distance estimate. Subtask B is the vehicle control task. An automatic transmission is assumed. The relative weightings of the cues are then calculated. Note that the creation of separate subtasks within the overall left turn task allows a more accurate assessment of the relative contribution of their respective cues.

Subtask A—Time of Collision and Distance to Cross

The object looming cue or rate of change in the angular size of the oncoming vehicle is estimated here based on the compiled data from Landwehr et al. (2013). For the purpose of this example, the original data (JNDs) have been converted to a common standard of JND_{99}. The use of this more conservative estimate is based on the belief that laboratory-based discriminatory thresholds are likely to overestimate the discriminative ability of the observers and thus the reliability of cueing in these studies. A more conservative estimate of the observer's discriminative ability is prudent when these data are generalized to the environment of actual vehicle operations, which impose a greater burden of noise on the operator's perceptual system. Using this more conservative criterion for the TTC of objects approaching drivers at a constant rate calculated by Landwehr and associates, relative changes in the object angular size have a discrimination threshold or JND_{99} of 0.09.

The second important cue in Subtask A is the distance that the cue used to judge how far the vehicle must travel to enter the intersecting road and clear the opposing lanes. The relative density cue weighting of Cutting and Vishton (1995) at the 10 m distance is used with a conversion again to the JND_{99} equivalent. A simplifying assumption is made that there are no objects of any kind in the intersecting road and thus no relative size cues are available.

Subtask B—Vehicle Control

Subtask B includes the cues essential to the control of the vehicle speed and direction.

The most significant cue is that of self-motion and heading changes provided by the optic flow of the visual scene. For the JND_{99} estimate here, data from Sun et al. (2003) for the vision-only condition were used.

The next cue listed is the cue that results from the vehicle movement associated with the turn itself. Data from Muller et al. (2013) for linear acceleration JND in the longitudinal axis were used. Once again, the JND data were converted to JND_{99}.

The remaining cues are torque and force cues from the steering wheel, accelerator, and brake pedals. For the steering wheel cue, force perception JND from the study of Newberry et al. (2007) was used. Pedal force JNDs were derived from the data of Southall (1985).

Once the relative weights are assigned to each cue, calculations can be conducted to determine what trade-offs that might be made in the design and their effects on

PFI_{des}. The PFI_{des} is the desired level of perceptual fidelity needed to support the task. It is the same as the relative weight calculation for each cue. As these relative weights are each calculated as a proportion of the sum of all cue weightings, full (100%) perceptual fidelity is achieved when all of the perceptual cues identified in Table 7.2 are provided by the simulator.

At this point, decisions on technology, cost, or other factors will be made and the inevitable trade-offs in design will take place. For the example here, it will be assumed that the customer has decided that the cost associated with providing the linear acceleration vestibular cueing (a motion platform) is not justified by the increment in vehicle control performance that might result. This reduces the PFI_{obt}/PFI_{des} to 86% for Subtask B because the relative weight of the acceleration cue was 0.14 or 14% of the relative weighting of all cues for this subtask.

As no change was made to any of the cues for Subtask A, the PFI_{obt} value is the same as the PFI_{des} and perceptual fidelity remains at 100% for this subtask. Full perceptual support is desired for the subtask in the vehicle simulator and expected operator performance should be comparable to that which would occur in the real-world vehicle operation under similar conditions. In contrast, the operator performance would be somewhat impaired by the loss of the linear acceleration cue in Subtask B. However, the customer is willing to accept this deficiency in performance for this particular subtask for the large initial and life cycle cost savings achieved.

The aforementioned example is only for a single task in what may be an extensive task requirements document. Only when all of the tasks are decomposed and relative weighting of the cues are calculated, can the design decisions can be made. This is due to the possibility that a particular design component may have an impact on multiple tasks. Moreover, the elimination of a design component and the perceptual cues that it provides may have different effects for different tasks. A linear deceleration cue produced by the motion platform that was eliminated in the aforementioned example for acceleration cueing may have a more significant effect on, for example, the deceleration cueing associated with emergency braking.

A particular measure of perceptual fidelity should be evaluated in the context of the use to which a vehicle simulator is to be put. Both costs, including costs associated with additional training and evaluation in the real vehicle, as well as benefits need to be examined. Analyses should also include the impact of eliminating a design component on the task proficiency in simulator performance. If some cues necessary to support a task are eliminated due to design changes, performance proficiency in the simulator may not be attainable. In this case, simulator training to proficiency may not be possible no matter how many training sessions are provided. The design component change in this case may not be cost-effective. Of course, the customer may not be interested in training an operator to proficiency in the simulator preferring that proficiency be achieved in the real vehicle. In this case, design component changes that result in reductions in perceptual fidelity of the vehicle simulator may be cost-effective.

Optimizing perceptual fidelity would seem to be self-evident for research simulator applications. However, it would depend on the research question that is involved. For example, the research team may only be interested in testing issues which involve visual perception cues and not the nonvisual cues that may accompany them. However, the research team must always be concerned about how perceptual fidelity might affect the ability to generalize the results of their study to the real world. Reductions in the perceptual fidelity of the research simulator may have an effect on the ability to generalize findings from their device to real vehicle operating environments.

DESIGN EVALUATION

As noted in Chapter 1 of this book, behavioral validation of the simulator design will eventually occur whether it is planned or not. Vehicle simulators used in training, testing, and research involve human operators engaged in meaningful tasks that usually have known performance criteria. Evaluation of operator behavior should, as noted earlier, include a sufficient representative sample of experienced vehicle operators. This allows testing of the functionality of the device across a wide range of representative operator performance levels.

Human–vehicle measures such as speed control for driving simulators and attitude control for flight simulators should be recorded and compared to archival data for this particular user population and vehicle type. Archival performance data can be from either real-world vehicle operations or those performed in a vehicle surrogate such as a research simulator. However, the surrogate must meet a high level of perceptual fidelity itself. It is understood that the instructor or other test personnel ratings of operators are often substituted for objective measures. However, these measures lack the reliability and validity necessary for definitive measurement of operator behavior.

Where possible, workload ratings should be obtained from the operators during the device evaluation particularly for those tasks deemed of particular importance. Workload ratings are sensitive to aspects of behavior not necessarily reflected in performance data such as the degree of mental and physical effort required to perform a particular task. Thus, although the performance data might indicate a high level of performance in executing a particular task, workload ratings may indicate that abnormal amounts of mental and physical effort may be needed to perform the task. Adaptation to some degree is involved when an individual moves from the real to the simulated vehicle. However, if perceptual fidelity is high, the amount of adaptation should be small. However, if excessive demands are being placed on the operator during device evaluation it is likely that relearning of a skill or behavior rather than simple adaptation is occurring. This will be reflected in unusually high workload ratings where such ratings would not be normally expected. Adaptations are relatively minor and are short-lived variations in behavior in response to new environments. Relearning behavior in a simulator reflects potentially serious shortcomings in the perceptual fidelity of the device. Adaptation, as it might be said, is measured in minutes, whereas relearning of a skill is measured in hours or days.

SUMMARY

Implications for design of vehicle simulators when perceptual fidelity is the design goal are discussed. The need to identify sensory perceptual limitations and capabilities of the intended user population is discussed. The design process is then elucidated with particular attention to the importance of task requirements definition and task decomposition. The latter is incorporated in task analyses that require detailed analyses of perceptual cues. The procedure required for calculating the relative weights of perceptual cues is described and the means of calculating both the PFI_{des} and the obtained perceptual fidelity index PFI_{obt} are presented. Examples of how design trade-offs can affect PFI_{des} and their effect on simulator value with regard to operator proficiency are discussed.

REFERENCES

Agrawal, Y., Carey, J.P., Della Santina, C.C., Schubert, M.C., and Minor, L.B. 2009. Disorders of balance and vestibular function in U.S. *JAMA Internal Medicine*, 169, 938–944.

Benson, A.J. 1990. Sensory functions and limitations of the vestibular system. In R. Warren and A.H. Wertheim (Eds.), *Perception and Control of Self Motion*. Hillsdale, NJ: Lawrence Erlbaum Associates.

Clark, B. and Stewart, J. 1972. The power law for the perception of rotation by airline pilots. *Perception and Psychophysics*, 11, 433–436.

Cohen, J. 1988. *Statistical Power Analysis for the Behavioral Sciences*. Hillsdale, NJ: Lawrence Erlbaum Associates.

Cutting, J.E. and Vishton, P.M. 1995. Perceiving layout and knowing distances: The integration, relative potency, and contextual use of different information about depth. In W. Epstein and S. Rogers (Eds.), *Perception of Space and Motion: Handbook of Perception and Cognition*. pp. 69–117. New York: Academic Press.

Dodgson, N.A. 2004. Variation and extrema of human interpupillary distance. In J. Woods, J.O. Merritt, S.A. Benon, and M.T. Bolas (Eds.), *Proceedings of the SPIE, Vol. 5291, Stereoscopic Displays and Virtual Reality Systems XI*. January 19–20, San Jose, CA.

Goble, D.J., Mousigan, M.A., and Brown, S.H. 2012. Compromised encoding of proprioceptively determined joint angles in older adults: The role of working memory and attentional load. *Experimental Brain Research*, 216, 35–40.

Kirwan, B. and Ainsworth, L.K. 1992. *A Guide to Task Analysis*. London, UK: Taylor & Francis Group.

Landwehr, K., Baures, R., Oberfeld, D., and Hecht, H. 2013. Visual discrimination thresholds for time to arrival. *Attention, Perception, and Psychophysics*, 75, 1465–1472.

Lopez-Escamez, J.A., Carey, J., Chung, W.-H., Goebel, J.A., Magnusson, M., Mandalà, M., Newman-Toker, D.E., Strupp, M., and Suzuki, M. 2015. Diagnostic criteria for Menière's disease. *Journal of Vestibular Research: Equilibrium & Orientation*, 25, 1–7.

Miller, R.B. 1953. *A Method for Man-Machine Task Analysis*. Wright-Patterson Air Force Base, OH. WADC Tech Report No. 53,137.

Muller, T., Hajek, H., Radic-Weisenfeld, L., and Bengler, K. 2013. Can you feel the difference? The just-noticeable difference of longitudinal acceleration. *Proceedings of the Human Factors and Ergonomics Society*, pp. 1219–1223. Thousand Oaks, CA: SAGE Publications.

Newberry, A.C., Griffin, M.J., and Dowson, M. 2007. Driver perception of steering feel. *Journal of Automotive Engineering*, 221, 405–415.

Shields, R.W. 2010. Peripheral Neuropathy. *Disease Management*. Cleveland Cline for Continuing Education. Retrieved from www.clevelandclinicmeded.com (accessed April 24, 2017).

Southall, D. 1985. Discrimination of clutch pedal resistances. *Ergonomics*, 28, 1311–1317.

Sun, H.J., Lee, A.J., and Campos, J.L. 2003. Multisensory integration in speed estimation during self-motion. *Cyberpsychology and Behavior*, 6, 509–518.

Index

Note: Page numbers followed by f and t refer to figures and tables respectively.